Nanotechnologies in the Conservation of Cultural Heritage

Piero Baglioni • David Chelazzi • Rodorico Giorgi

Nanotechnologies in the Conservation of Cultural Heritage

A compendium of materials and techniques

Springer

Piero Baglioni
Department of Chemistry and CSGI
University of Florence
Florence
Italy

Rodorico Giorgi
Department of Chemistry and CSGI
University of Florence
Florence
Italy

David Chelazzi
Department of Chemistry and CSGI
University of Florence
Florence
Italy

ISBN 978-94-017-9302-5 ISBN 978-94-017-9303-2 (eBook)
DOI 10.1007/978-94-017-9303-2
Springer Dordrecht Heidelberg New York London

Library of Congress Control Number: 2014952024

Printed on acid-free paper

Springer is part of Springer Science+Business Media (www.springer.com)

Preface

The idea of writing this book, mainly aimed at practitioners in the field of Cultural Heritage preservation, came from continuous interaction with conservators and their requirement for practical guidelines on the use of nanomaterials for conservation. Accordingly, this compendium deals with both the fundamental principles and practical usage of different innovative nanomaterials for several classes of common movable and immovable artistic substrates. The ideal audience comprises all the conservation end-users, such as freelance conservators and restorers, personnel and experts of conservation institutions and museums, and also students of restoration schools (including universities and colleges that adopted conservation programs) and scientists who are willing to approach the conservation of artifacts, historical objects and works of art.

The methodologies developed by the authors in the last decades are described with a clear approach that is accessible also to readers and end-users that might come from humanistic or artistic areas and are not familiar with chemistry and advanced conservation systems/technology, while maintaining scientific rigor and providing references to advanced material science. The reported methodologies for the consolidation, cleaning and deacidification of artifacts, cover a significant and representative part of the case studies met in Cultural Heritage preservation. The approach described is based on the use of advanced systems (derived from colloid and soft matter science) applied to practical restoration needs, in fact most of the methods reported in this compendium have already been used in restoration facilities, workshops and ateliers worldwide.

The conservation systems described in this compendium represent a significant improvement on traditional conservation methodologies because the proposed advanced systems exhibit enhanced properties, namely:

- High physico-chemical compatibility with the constituents of works of art, i.e. the application of the advanced systems minimizes or completely avoids any alteration of the original physical and chemical properties of the artistic/historical substrates.
- The proposed nanomaterials are either non-toxic or they exhibit a significantly reduced toxicity as compared to traditional restoration materials such as pure solvents or solvent blends.

- The use of advanced nanomaterials allows greater control of the restoration intervention, for instance highly controlled cleaning can be carried out using microemulsions and chemical hydrogels as compared to traditional cleaning methods.
- The innovative methodologies proposed in the compendium are feasible and reliable, and in many cases simpler and faster than traditional methods. In some cases methods are proposed where a gradual and slower approach is necessary to grant stability of the treated artifact also in the long term (as opposed to "quick" traditional interventions that might involve drawbacks, requiring later interventions).

Even though the proposed application of the advanced systems is feasible, the development of these systems derives from scientific knowledge of sometimes complex concepts. Moreover, it must be noticed that some of these systems are self-assembly under a range of conditions (temperature, pressure, composition, etc.), and their production requires fine knowledge of phase behaviors. Therefore, it is highly inadvisable to modify the original systems through empirical and trial-and-error procedures, as the composition and practical use suggested by the authors of this compendium already grant the best conditions for optimal efficiency. The arbitrary modification of these systems involves the risk of reducing their effectiveness, and in some cases this might result in potentially detrimental effects on the treated works of art. Conservators and end-users are thus encouraged to use the proposed materials and methodologies as they have been described in the dedicated chapters, and to contact the authors of this compendium for further guidance and recommendations on the use of these formulations. We believe, in fact, that the synergistic cooperation between different disciplines and skills is a fundamental step in the conservation of cultural heritage.

The compendium contents are structured as follows: first, an overview is provided about the main nanomaterials developed in the last 30 years (dispersions of nanoparticles, micellar solutions, microemulsions and gels), explaining their main features and applicability. Then, specific chapters are dedicated to artistic/historical substrates (e.g. wall and easel paintings, stone, paper, canvas and wood), dealing with the main artifacts degradation processes and then discussing practical applications and guidelines for the cleaning, consolidation, or deacidification of works of art.

Throughout the writing of this compendium we have been in contact with colleagues, both colloid/soft matter scientists and conservators, who are gratefully acknowledged for discussion and support as we were engaged in the effort of describing simple yet comprehensive applicative methodologies. We also wish to acknowledge the European Union (which has financed the FP7 project "Nanoforart") for partly supporting some of the activities that ultimately led to the development of practical protocols for the use of nanomaterials in conservation of Cultural Heritage.

Florence, July the 18th, 2014

Piero Baglioni
David Chelazzi
Rodorico Giorgi

Contents

About the Authors

Piero Baglioni is a full professor of Physical Chemistry and lecturer of Physical Chemistry at the Department of Chemistry of the University of Florence. He is the Director of the Center for Colloid and Surface Science (Consorzio Interuniversitario per lo Sviluppo dei Sistemi a Grande Interfase—CSGI); he is in the Advisory Board of several international journals, International organizations, and member of several national and international institutions, industries, and learned societies (e.g the European Academy of Science; the Royal Academy of Art and Science in Göteborg, Sweden; the American Chemical Society; the European Colloids and Interface Society; the European Society for Neutron Spectroscopy; Accademia delle Arti del Disegno—the oldest Academy in Italy). Piero Baglioni is author of about 400 publications on books and journals and 26 patents (5 of which in connection with art restoration). He was recipient of international prizes for his contribution to Conservation of Cultural Heritage and to the synthesis of nucleolipids, such as Lifetime achievements for the contribution to Colloids and Interface Science (JCIS), the Caballero Aguila (The most prestigious recognition from National agency for conservation in Mexico), the Gran Prix for Innovation Award and from the European Colloids and Interface Society (ECIS). The methods for the conservation of artistic and historical objects introduced by his group have been widely acknowledged.

David Chelazzi received his Master's degree in Chemistry at the University of Florence in 2003, with a thesis work developed at the European Laboratory for Non-Linear Spectroscopy (L.E.N.S.). In 2007 he received his PhD on Science for Cultural Heritage Conservation at CSGI, on the development of physico-chemical methodologies for the conservation of paper and wood works of art. Post-doc experience included a scholarship at the "Centre de Restauration des Musées de France" (C2RMF) in Paris (Palais du Louvre), on the aging and removal of adhesives used for the lining of paintings.

As of 2010 he is a Research Fellow at CSGI—Department of Chemistry of the University of Florence, his research work focuses on the development of nanomaterials for the conservation and preservation of movable and immovable works of art. He is the author of 30 publications in the field of conservation of cultural heritage materials.

Rodorico Giorgi received his degree in Chemistry (physical-chemistry curriculum) in 1996, at the University of Florence, and in 2000 his PhD in Science for the Conservation of Cultural Heritage at the same University, with Prof. Piero Baglioni as Advisor. He is currently research fellow at the Department of Chemistry of the University of Florence and CSGI, national Center for Colloid and Surface Science. Giorgi's main research interests are in physico-chemical characterization of materials, investigation of degradation processes and development of nanotechnology for the conservation of different works of art materials, such as wall and canvas paintings, stone, paper, parchment, and archaeological wood. Giorgi is author of about 100 publications in the field of nanoscience application to cultural heritage conservation.

Chapter 1
Innovative Nanomaterials: Principles, Availability and Scopes

Abstract The collaboration between scientists and conservators is fundamental for achieving reliable and durable results in the conservation of works of art. The contribution of scientists lies in the development of both advanced diagnostic techniques and cutting-edge materials for the restoration of artifacts. Colloid and materials science have provided over the last decades a palette of tools for the cleaning, consolidation and pH control of artistic and historical substrates. Nanostructured materials such as microemulsions, micellar solutions, dispersions of alkaline nanoparticles and chemical gels, can be used to effectively counteract the degradation processes without altering the physico–chemical properties of the treated works of art, and to minimize or completely avoid drawbacks. This chapter provides an overview of the principles and scopes that underlie the use of nanomaterials in conservation science, considering the advantages with respect to traditional restoration materials, and the availability of these innovative tools.

1.1 Nanotechnology for Cultural Heritage Preservation

Cultural Heritage is an invaluable patrimony of society, embracing virtually all the artifacts, works of art, objects and intangible attributes that convey artistic, historical or anthropological values. The preservation of such patrimony is the only way to effectively transfer it to future generations, in order to continue the intellectual progress of society while conserving the ancient and modern cultural traditions that characterize our world. Besides intellectual and aesthetic aspects, the preservation of Cultural Heritage allows its valorization and exploitation, with considerable economic advantages. Given its importance, it is not surprising that Cultural Heritage has gathered in the last decades the attention of different professional characters that provided approaches to address numerous conservation issues. In fact, the variety of degradation phenomena that affect works of art

P. Baglioni et al., *Nanotechnologies in the Conservation of Cultural Heritage*,
DOI 10.1007/978-94-017-9303-2_1

mirrors the vast array of materials that have been used by mankind since early ages. Therefore both conservators and scientists are involved in finding effective solutions to counteract aging processes due to the action of light, temperature, relative humidity and microorganisms, chemical degradation and physical erosion, or to anthropic causes such as industrial pollution, vandalism, or the mere handling of artifacts. Moreover, restoration interventions can prove—and have often proven—detrimental in the long term whenever scientific criteria are not followed. Based on the experience acquired in the past decades, the use of products that exhibit as much as possible the same physico–chemical properties of the treated artistic or historical substrates (i.e. "compatible" materials) has been highlighted as a valid principle to grant the durability of treatments and to minimize drawbacks. For instance, the treatment of carbonate-based wall paintings with low-compatibility materials such as synthetic organic coatings and adhesives can lead to the alteration and degradation of the painted surface, and compatible inorganic materials have been successfully proposed as an alternative for the consolidation of these works of art, see for instance works by Ambrosi et al. (2001), Giorgi et al. (2010a, b), and Chelazzi et al. (2013).

A multidisciplinary approach to conservation issues is the key for a successful intervention, and the cooperation between scientists, conservators, art historians etc. is fundamental for the refinement of restoration materials and techniques.

In this framework, the role of science is comparable to that of medicine: gathering information on the "anatomy" and "physiology" of the "patient" (i.e. the composition and physico–chemical properties of the artifact) is the first task that is pursued through the use of literature, previous knowledge and both advanced diagnostic and computational techniques that are also fundamental in determining the nature of the "disease" (the degradation processes affecting the artifact) and in checking the effectiveness of restoration interventions (Bianchin et al. 2009; Fantacci et al. 2010). Finally, a "cure" must be found, meaning that tools and materials are to be developed to counteract, stop, and ideally revert the degradation process. Regarding the latter task, which is the focus of this Compendium, materials science has provided a fundamental contribution, and in particular colloids science and nanosciences have emerged in the last four decades as fields of paramount importance, being the source of concepts and tools that have improved dramatically the effectiveness, reliability and durability of restoration interventions (Baglioni and Chelazzi 2013; Baglioni et al. 2013; Baglioni and Giorgi 2006).

The aim of this chapter is to introduce the main innovative conservation materials developed by colloids and material science, namely nanoparticles, nanostructured cleaning fluids and gels. Each section will highlight the principles that underlie the use of such materials, the availability for end users, and the scopes of each class of materials. Then, Chaps. 2–5 will provide practical information on the use of nanoparticles, fluids and gels to address conservation problematics, each chapter dealing with a specific task (e.g. cleaning of easel paintings, consolidation of wall paintings, deacidification of paper, etc.).

Edge length	4	2	1
Total surface area	96	192	384
Total volume	64	64	64
Surface-to-volume ratio	1.5	3	6

Fig. 1.1 Total surface area, volume and surface-to-volume ratio of a (model) cubic mass as it is divided into sub-units. Surface-to-volume ratio increases while total volume remains constant. (Image by Michele Baglioni)

1.2 Nanoparticles

In conservation science, the application of solid particles dispersed in a medium can be preferred to either aqueous or organic solvents solutions for different reasons. The solubility of a consolidant in water might be too low to allow its effective use as an aqueous solution, while the same material could be applied in larger quantities as a solid stably dispersed in an organic solvent. Deacidifying agents such as calcium hydroxide ($Ca(OH)_2$) nanoparticles dispersions can be deposited onto paper fibers, where they neutralize acidity (Giorgi et al. 2002). The particles excess reacts with atmospheric CO_2 and turns into calcium carbonate ($CaCO_3$) that being a milder alkaline compound is not detrimental to aged (oxidized) cellulose, and acts as a solid buffer against recurring acidity. On the other hand, the use of alkaline solutions delivers all at once a large amount of mobile hydroxide ions that could be harmful to cellulose.

In several cases, when dispersions of particles are used for conservation tasks the effectiveness of the application increases as the size of the particles is reduced down to the nanoscale.

Materials are defined as "nanostructured" if they exhibit at least one dimension in the range of 10^{-9} m. Accordingly, nanomaterials can be bidimensional (sheets with nanometric thickness), monodimensional (rods, tubes, wires or cylindrical micelles with nanometric diameter) or zerodimensional.

As the size of particles is reduced, the surface area per unit volume increases, as illustrated in Fig. 1.1. The material's reactivity is consequently enhanced, since more active surface will be disposable for reactions and transformations to take place. In other words, the interface between the particles and the external environment becomes larger if the same mass of matter is divided into finer particles. The reactivity enhancement is important in practical applications, for instance when calcium hydroxide nanoparticles are used for the deacidification of paper as mentioned above. An increased reactivity makes the particles better deacidifying agents, and favors the transformation of $Ca(OH)_2$ into the milder alkaline buffer $CaCO_3$.

Size reduction is also important to ease the dispersion of solid particles into carrier solvents. Dispersions can then be easily brushed, sprayed or dripped onto artistic surfaces. Stable formulations do not require the use of stabilizers. For instance, reasonably stable dispersions of calcium and magnesium hydroxides nanoparticles can be prepared using short chain alcohols such as ethanol and propanol (see Fig. 1.2) without the need of surfactants that might remain as residues on the surfaces treated with the dispersions, after the evaporation of solvents.

Another advantage related to particle size decrease is to favor penetration through porous matrices, such as wall paintings, wood, surface coatings, etc. The well known Bookkeeper method (Preservation Technologies, L.P.) involves the use of particles of magnesium oxide, MgO, with a diameter slightly smaller than 1 μm, for the effective deacidification of paper, but the method is discouraged when paper porosity is low since in that case particles can not penetrate completely through the paper fibers, and deposits can be left as white hazes over the treated surfaces. Moreover several additives are present in the formulation to stabilize the dispersion. Magnesium hydroxide nanoparticles have been successfully used for paper deacidification (Poggi et al. 2010, 2011), and the small size of the particles increases penetration through the fibers, minimizing the risk of haze formation and favoring the homogeneous penetration and distribution of the deacidifying agent.

Nanosized particles are also useful to enhance the properties of organic–inorganic hybrid composites, which typically include a polymeric binding matrix and inorganic fillers (particles). Nanocomposites exhibit at least one component with nanometric size or nanostructuration, and show better performances than traditional filled polymeric matrices in terms of mechanical properties, chemical resistance, protection against UV radiation, etc. These materials can be used for conservation purposes, for instance polymer–silica nanoparticles composite films show increased hydrophobicity with respect to poly(methyl methacrylate) coatings for the water protection of stone-based monuments (Manoudis et al. 2007), and a Klucel-TiO_2 nanocomposite has been proposed for the protection of paper from UV radiation and from the formation of biofilms (Afsharpour et al. 2011).

The use of nanoparticles for the preservation of works of art will be detailed in Chap. 2 (consolidation of wall paintings and stone) and Chap. 5 (deacidification of paper, canvas, and wood), where the most recent advances and applications will be described, together with practical guidelines.

The synthesis of nanoparticles and the preparation of stable particles dispersions are normally carried out at scientific and specialized facilities, since both the required skills and equipment rule out the possibility of preparing homemade formulations in ateliers, restoration laboratories etc. However, several products for restoration are presently sold on the market.

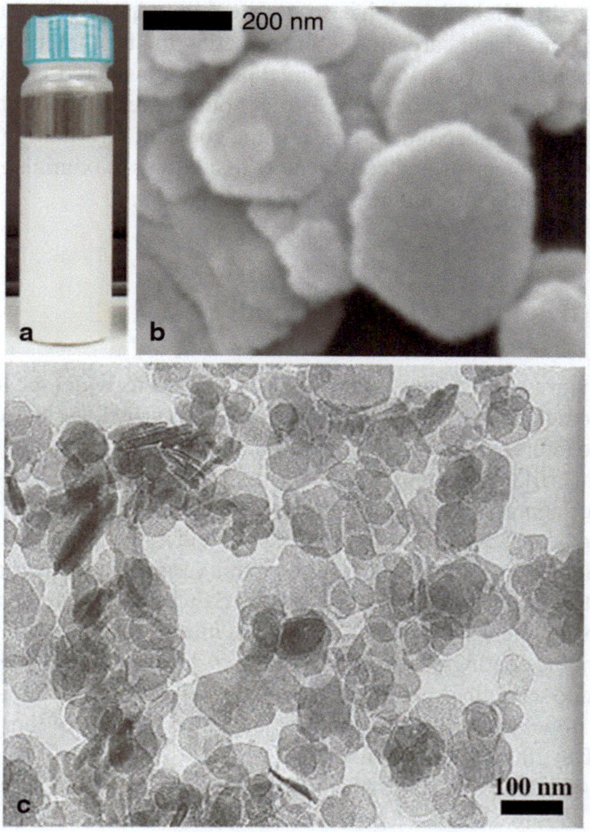

Fig. 1.2 **a** A dispersion of calcium hydroxide nanoparticles in 2-propanol. Reprinted with permission from Carretti et al (2013) Interactions between nanostructured calcium hydroxide and acrylate copolymers: implications in Cultural Heritage conservation. Langmuir 29:9881. Copyright 2013 American Chemical Society. **b** Scanning electronic microscopy (SEM) image of calcium hydroxide ($Ca(OH)_2$) nanoparticles. Reproduced from Baglioni and Giorgi (2006) Soft and hard nanomaterials for restoration and conservation of Cultural Heritage. Soft Matter 2:293, with permission from The Royal Society of Chemistry. **c** Transmission electronic microscopy (TEM) image of magnesium hydroxide ($Mg(OH)_2$) nanoparticles. Reprinted with permission from Giorgi et al (2005) Nanoparticles of $Mg(OH)_2$: synthesis and application to paper conservation. Langmuir 21:8495. Copyright 2005 American Chemical Society.

The application of nanoparticles dispersions for the consolidation of stone and wall paintings, or for the deacidification of cellulose, can be relatively easy once the fundamental properties of the artistic substrate are known, as well as the environmental conditions experienced by the artifact during and after treatment. Years of experience and collaboration with conservators on real case studies have led the authors of this Compendium to formulate practical guidelines for the use of the nanomaterials that they have designed. For what concerns the application of nanoparticles, such guidelines will be described in Sects. 2.4 and 5.4. However, it is important to highlight that each case study exhibits specific features and issues, which must be considered carefully before selecting the best intervention protocol. Therefore, it is imperative that these guidelines be considered critically case-by-case, rather than applied mechanically. Throughout any restoration intervention that involves the use of advanced materials, it is the constant information and expertise exchange between scientists and conservators that allows maximizing the effectiveness of protocols while limiting, or avoiding, drawbacks.

1.3 Nanostructured Cleaning Fluids

Cleaning of a work of art can be a very delicate task. In its wider meaning, cleaning involves the removal of any undesired material from artistic or historical surfaces. In practice, the intervention must be carried out in a controlled way without damaging the original artifact due to mechanical stress or any other process such as swelling, leaching of components, discoloration, etc.

Materials to be removed come in great variety and range from dirt, grime and soil to natural and synthetic coatings, adhesives and varnishes, which can darken and degrade over time so that they produce aesthetic alteration or even physico–chemical degradation of the artistic substrate. As a matter of fact the application of synthetic hydrophobic coatings, for instance poly acrylates and vinyl-acetates, strongly alter the physico–chemical properties of substrates such as carbonate-based wall paintings and stone, resulting in medium or long-term damage of the artistic surface (Carretti and Dei 2004; Giorgi et al. 2010a, b).

The extent to which undesired layers are removed can be a matter of debate and depends from case to case based on historic, aesthetic or ethic factors. For instance, sometimes a "patina" is left on top of artistic surfaces since it is considered as a historical part itself. Recently, an "overcleaning" issue arose concerning the restoration of Leonardo's painting "The Virgin and Child With Saint Anne" (Alberge 2011).

Through the ages, a vast array of cleaning materials has been used, such as soaps (from fats and oils), alimentary products (vinegar, wine, lemon juice), inorganic materials (potash solutions) and even biofluids like urine, bile or saliva, which is nowadays synthetically prepared and used by conservators (Wolbers 2000). Some of these fluids (bile, saliva, soaps) contain surfactants, which have been used since ancient times and have been widely studied and diffused by synthetic industry starting from the twentieth century. Nowadays surfactants are involved in everyday life ranging from detergency to food chemistry, pharmaceuticals, etc., Branches of modern advanced chemistry and physical chemistry such as materials, colloid and nanosciences have carried out intense research on both theoretical and applicative aspects related to soft matter, such as binary or ternary systems composed of water, surfactants and other additives (e.g. solvents).

The end of the 1980s saw the first application of nanostructured fluid systems based on surfactants for the removal of wax spots from the painted surface of Italian Renaissance wall paintings in Florence (Borgioli et al. 1995; Baglioni et al. 2013).

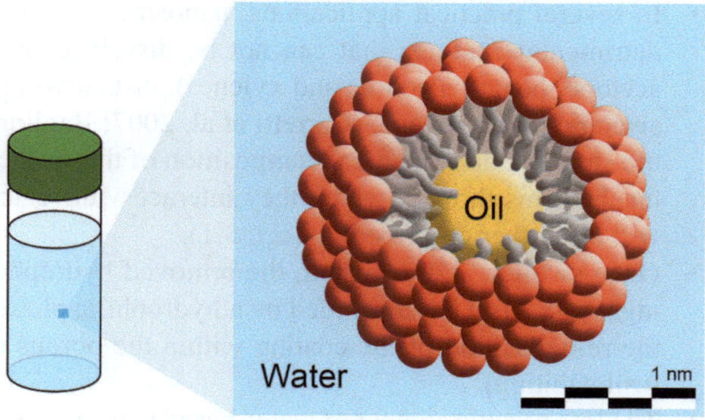

Fig. 1.3 An oil-in-water (o/w) microemulsion, where organic solvents are dispersed into nano-sized surfactant micelles surrounded by a continuous water phase. (Image by Michele Baglioni)

Starting from that pioneering study, several systems have been prepared characterized and successfully applied for the removal of undesired materials from artistic surfaces. Numerous examples are reported by the literature, see for instance Baglioni et al. (2011) and Carretti et al. (2003, 2007), making nanostructured fluids one of the most appealing advanced cleaning tools for Cultural Heritage conservation, together with gels (see Sect. 1.4), laser technology (Nevin et al. 2007; Siano and Salimbeni 2010; Pouli et al. 2010), and the promising "biocleaning" based on the use of microorganisms (Cappitelli et al. 2006; Alfano et al. 2011).

One typical class of nanostructured fluids used for artifacts cleaning are oil-in-water (o/w) microemulsions, where an organic solvent is dispersed into nano-sized surfactant micelles surrounded by a continuous water phase (see Fig. 1.3). These systems are thermodynamically stable and optically transparent; they exhibit a high water content (75–99 %) and a reduced organic content (from less than 0.5 % to about 15 % including both solvents and surfactants), nonetheless they have proven highly effective in swelling and removing detrimental coatings from works of art. In fact microemulsions and other surfactant-based fluids have emerged as a valid alternative to the use of organic solvents, which involves several drawbacks: first, the action of free solvents is not completely controllable and might lead to the swelling and solubilization of the artifact's organic components (e.g. binders, organic additives); secondly, coatings and grime dissolved by solvents are then transported through porous matrices and deposited in the artifact's pores; finally, but most importantly, many solvents are toxic and conservators often work in scarcely aerated environments. The use of gelled solvents can limit these drawbacks (see Sect. 1.4).

Compared to these limitations, the use of nanostructured cleaning fluids involves the following advantages:

- The processes that lead to the swelling (and/or solubilization) and removal of substances (polymers, wax, etc.) include interaction and matter exchange at the interface between the nano-sized containers (micelles), the aqueous continuous phase and the surface of the layer to be removed. The micelles exhibit a large surface area; therefore the extended interface maximizes the interaction with the detrimental layer and its swelling/solubilization.

- In several practical applications, nanostructured fluids can remove effectively detrimental coatings that can not be dissolved by conventional solvents (e.g. acetone, benzyl alcohols and xylenes), or that would require the use of aggressive and toxic solvents (Carretti et al. 2007; Baglioni et al 2012). This is due to the structure and chemical composition of the surfactant-based fluids, and to the mechanisms through which they interact with organic coatings (the interaction will be further described in Sect. 3.1).
- Once swollen and detached, the removed hydrophobic material (organic coatings, wax, etc.) is surrounded by a hydrophilic phase (aqueous phase) that limits the re-deposition of the coating within the porous matrix of the substrate (e.g. wall painting).
- The low organic content of water-based nanostructured cleaning fluids depresses the health risk and the environmental impact involved in the cleaning intervention.
- Microemulsions are thermodynamically stable systems, which means that they can be used in different environmental conditions without causing the formation of two separated macroscopic phases (organic and aqueous). In stable systems, the organic content is dispersed in water (see Fig. 1.3), and the system is macroscopically monophasic (optically clear).

As mentioned above, microemulsions were initially designed for the cleaning of wall paintings. However, the application of water-based cleaning fluids on water-sensitive substrates (e.g. paper and canvas) is not only possible but also very effective when the fluids are confined in hydrogels, which allow controlling the fluid release and the cleaning action so that no detrimental effects, such as fiber swelling, is induced on the treated surface (see Sect. 1.4).

The literature reports several case studies where specific formulations of microemulsions and micellar solutions have been used for the removal of some of the main classes of synthetic coatings that have been applied on artifacts starting from the second half of the twentieth century. Sections 3.2, 4.2, 3.4 and 4.4 describe formulations and application procedures for the cleaning of wall paintings, stone and easel paintings. As previously mentioned for nanoparticles, it is important to notice that such guidelines must be adopted critically, and the cooperation between the nanostructured fluids designers (scientists) and end users (conservators) is essential to achieve the best results.

1.4 Gels

As described in Sect. 1.3, the use of solvents involves several drawbacks mainly related to safety risks and to the possibility of altering the artifact during cleaning. The ideal intervention should be fully selective, which means that removal of undesired layers must be carried out without affecting the original artistic materials either chemically or physically. One of the best methods to achieve such task is to confine solvents in a matrix that releases them gradually onto the substrate.

Over the last few decades, conservators have adopted different tools, including modified natural products such as cellulose ethers (e.g. Kucel®, Tylose), or synthetic polymers such as polyacrylic acids (e.g. Carbopol®). These materials can be used to thicken solvents, so to limit uncontrolled penetration through porous substrates.

In the late 1980s and early 1990s Richard Wolbers developed the so-called "solvent gels", which are obtained dispersing polyacrylic acid in a solvent and then adding weakly basic non-ionic surfactants like Ethomeen C12® and C25®. The bases cause deprotonation of carboxylic functions in the acid chains that unfold and form an extended 3D network, which confines the solvent (Wolbers et al. 1988; Wolbers 2000). Solvent gels are still one of the most used cleaning tools due to their effectiveness and versatility; in fact they can also be used to control the action of detergents and enzymes.

Both cellulose ethers and polyacrylic acid-based gel-like networks are built by molecules cross-linked through non-covalent interactions, such as dipole–dipole interactions and hydrogen bonding; therefore they are classified as "physical gels" as opposed to networks where cross-links are covalent bonds ("chemical gels"). The main drawback associated with physical gels is that they leave solid residues on the treated surfaces (Stulik et al. 2004). The long-term effects of these residues have not been completely clarified yet, and the common procedure to remove them is to use cotton swabs and solvents, which eventually brings back the aforementioned issues. Moreover, residual degradation products of Ethomeen (especially C25®) evaporate only partially from the painted layer, and their effects on the long-term are still to be fully characterized. A study by Burnstock and White (2000) indicated that, in regard to the effects of aging of combined samples of di- and triterpenoid resins, Ethomeen C12® has a synergetic influence on degradation processes, and the surfactant degradation products include amine N-oxides, which gave rise to concern about long-term contact with the resin or painted media.

Owing to these issues, scientific research has focused in the last decade on the formulation of alternative systems that can be easily and fully removed from the painted layer, minimizing or completely avoiding the presence of solid residues.

Polysaccharide-based gums have been recently considered for the cleaning of artistic and historical surfaces. For instance, agar (or agar–agar) is a mixture of agarose and agaropectin, and it is sold as powder that can be mixed with water, obtaining a gel whose porosity depends on the concentration of agarose. This gel can support solutions or water-based systems such as microemulsions for the treatment of porous substrates (Campani et al. 2007; Gorel 2010). Another example is gellan gum (Phytagel®), a microbial exopolysaccharide that can be gelled and used for the controlled cleaning of water-sensitive substrates such as works of art on paper (Iannuccelli and Sotgiu 2010).

Chemical hydrogels are another class of materials that have recently found application in the cleaning of artifacts, owing to characteristic features that make them advantageous with respect to other formulations. Domingues et al. (2013) developed semi-interpenetrating p(HEMA)/PVP networks where free chains of poly(vinylpyrrolidone)—PVP—are embedded into a network formed by poly(2-hydroxyethyl methacrylate)—p(HEMA). These systems exhibit the mechanical

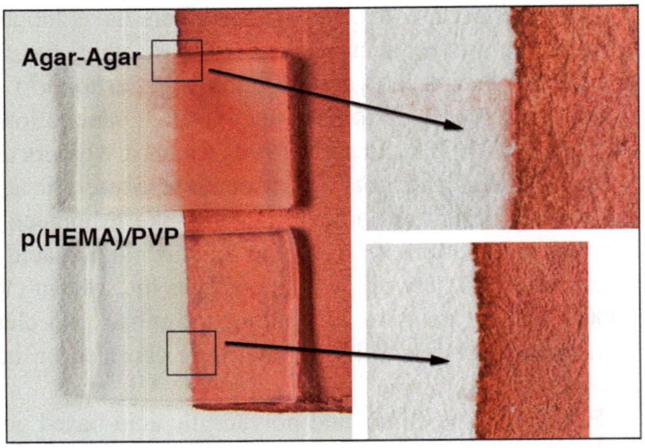

Fig. 1.4 Five minutes application of agar–agar (2 % w/w) and semi-IPN p(HEMA)/PVP hydrogels on paper painted with brazil-wood ink. (Reprinted with permission from Domingues et al. (2013) Innovative hydrogels based on semi-interpenetrating p(HEMA)/PVP networks for the cleaning of water-sensitive cultural heritage artifacts. Langmuir 29:2746. Copyright 2013 American Chemical Society)

strength of p(HEMA) combined with the hydrophilicity of PVP. In particular, both the p(HEMA)/PVP ratio and the amount of water used during the preparation steps affect the final properties of the gels, which can be tuned to obtain the desired properties, i.e. good adhesion to the substrate, ideal retention/release of the detergent system (water, o/w microemulsions, micellar solutions etc.), confinement of the cleaning action to the contact area between the gel and the artistic substrate.

Literature reports the size of both the macropores and of the mesh of p(HEMA)/PVP gels (Domingues et al. 2013). The mesh size intrinsically provides a measure of the average microporosity of the gel network, i.e. of the nano-sized pores. The average porosity values depend on the gel formulation, for instance a particular formulation (named H50) has macropores of 5–15 μm and a mesh of 2.5 nm.

The p(HEMA)/PVP gels are highly water-retentive, which is a crucial feature for granting controlled cleaning. In fact, the water release of these systems is strongly reduced with respect to AgarArt® (agar–agar gel) and Kelcogel® (gellan gel) prepared by dispersion of dry powders in water (3 % w/w), as shown by Domingues et al. (2014).

p(HEMA)/PVP hydrogels have been tested on highly water-sensitive and scarcely cohered painted surfaces, such as *tempera magra* paintings on canvas, where pigments and colorants are mixed with the minimal amount of binder (animal glue) necessary to wet the pigment particles. The gels, loaded with water, proved effective in the gradual and controlled removal of grime trapped into the substrate surface pores, without any alteration of the painted layer (swelling, leaching). The p(HEMA)/PVP gels were also tested on paper samples painted with a water-soluble ink (see Fig. 1.4). As a matter of fact, the application of the highly retentive chemical hydrogel did not alter the surface and no diffusion of the colorant through the paper substrate was observed, indicating that the use of the gel avoided excessive wetting. The same work demonstrated that, owing to their mechanic properties, the p(HEMA)/PVP gels can be easily handled and applied as sheets on the artistic surface. The gels network is more cohesive than physical gels because it is formed by covalent bonds that link polymer chains. As a consequence, after the application

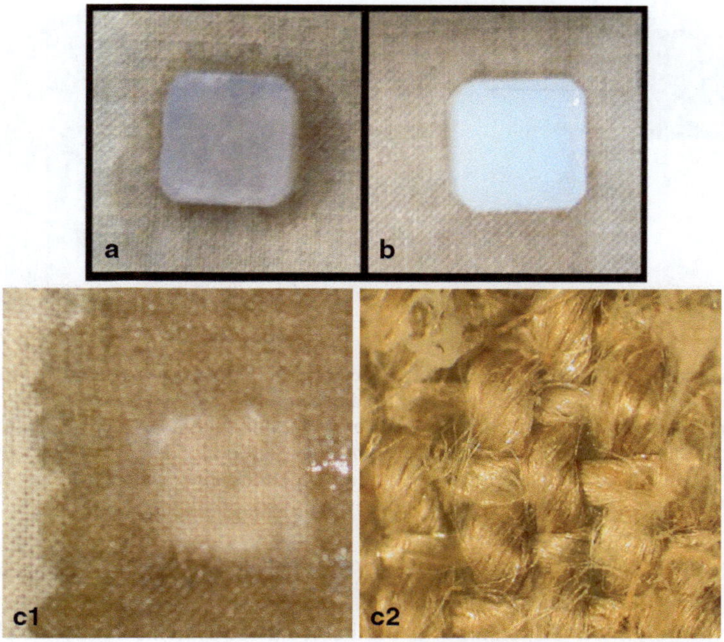

Fig. 1.5 **a**, **b** Acrylamide/bisacrylamide gels loaded with an oil-in-water microemulsion for the removal of lining adhesive from canvas. **b** The gel formulation can be tuned so to grant wettability of the fibers while avoiding the diffusion of water. **c1** Result of the cleaning test: the aged and darkened adhesive has been removed from the canvas. **c2** microscopy magnification (×10) of the same area. (Readapted with permission from Pizzorusso et al. (2012) Physicochemical characterization of acrylamide/bisacrylamide hydrogels and their application for the conservation of easel paintings. Langmuir 28:3952. Copyright 2012 American Chemical Society)

chemical gels can be removed without leaving detectable residues, as highlighted by Fourier Transform Infrared (FTIR) analysis.

A study by Pizzorusso et al. (2012) reported on the use of another class of chemical hydrogels, i.e. acrylamide/bisacrylamide gels, for the cleaning of easel paintings. The 3D network is produced by the radical polymerization of acrylamide monomer and a cross-linker, e.g. *N, N'*-methylene bisacrylamide. The macropores have size of 5–25 μm, while the mesh size is 7–9 nm. The resulting gels have a behavior that is similar to solids (see Fig. 1.5): they can be easily applied and removed from the treated surface without leaving any residues, as determined for the p(HEMA)/PVP gels. Acrylamide/bisacrylamide gels, loaded with a surfactant-based nanostructured fluid, were successfully used to remove detrimental synthetic adhesives from canvas. The gels grant good wettability of fibers while avoiding the diffusion of water through the substrate and along the plane of canvas surface, as shown in Fig. 1.5. By coupling the swelling of the adhesive layer by the fluid with gentle mechanical action, it was possible to achieve the selective cleaning of the canvas without causing alterations (swelling of the fibers).

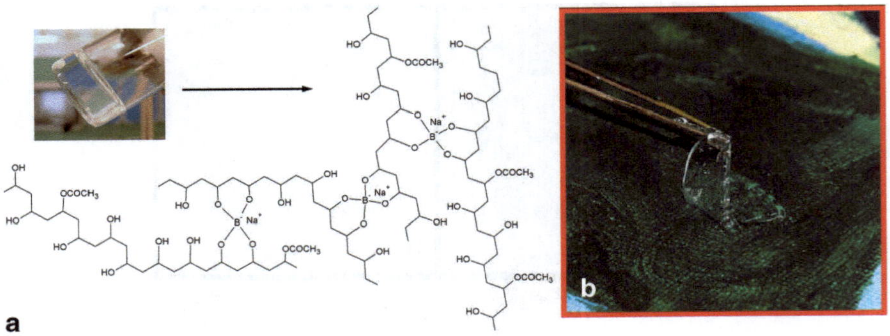

Fig. 1.6 **a** A High-viscosity polymeric dispersion (HVPD) formed by partially hydrolyzed poly(vinyl acetate), whose chains are cross-linked by borax. **b** The HVPD is safely peeled off from the surface of a canvas painting by means of tweezers. (Readapted from Natali I et al (2011) Structural and mechanical properties of "peelable" organoaqueous dispersions with partially hydrolyzed poly(vinyl acetate)-borate networks: applications to cleaning painted surfaces. Langmuir 27:13400. Copyright 2011 American Chemical Society.)

High-viscosity polymeric dispersions (HVPDs) are another class of useful materials for the selective removal of coatings from artistic surfaces (Natali et al. 2011). The term "polymeric dispersions" is used since these systems cannot be classified as gels based on their rheological behavior, even though they have a gel-like appearance (see Fig. 1.6). HVPDs can be formed by poly(vinyl alcohol) or by partially hydrolyzed poly(vinyl acetate), whose chains are cross-linked by borax. The resulting 3D networks are thermodynamically stable and exhibit features that are particularly appealing for the cleaning of works of art. In fact, these "gels" can adapt their shape so to maximize the contact with the artistic substrate. Depending on the degree of poly(vinyl acetate), the systems can be produced and loaded with water or organic solvents such as acetone, alcohols (ethanol, 1-propanol, 2-butanol, 1-pentanol), propylene carbonate, methyl ethyl ketone, *N*-methylpyrrolidone and cyclohexanone; therefore HVPDs can be used for the removal of different types of materials (Angelova et al. 2011; Carretti et al. 2009). Even more importantly, these systems exhibit a high elastic modulus; therefore it is possible to simply "peel" them from the treated surface after application, as shown in Fig. 1.6). For instance, a study by Carretti et al. (2010) highlighted that PVA-borax dispersions could be removed in one step (by peeling) without leaving any measurable amount of residues (Carretti et al. 2010).

Chemical gels and HVPDs will soon be available on the market, under the trademark Nanorestore Gel®. More details on the preparation and use of chemical gels and HVPDs will be provided in Chaps. 3, 4.

References

Afsharpour M, Rad FT, Maleikan H (2011) New cellulosic titanium dioxide nanocomposite as a protective coating for preserving paper-art-works. J Cult Herit 12:380–383

Alberge D (2011) The guardian, culture, art and design. http://www.theguardian.com/artanddesign/2011/dec/28/louvre-leonardo-overcleaned-art-experts. Accessed 10 Oct 2013

Alfano G, Lustrato G, Belli C et al (2011) The bioremoval of nitrate and sulphate alterations on artistic stonework: the case-study of Matera Cathedral after six years from the treatment. Int Biodeter Biodegr 65:1004–1011

Ambrosi M, Dei L, Giorgi R et al (2001) Colloidal particles of Ca(OH)2: properties and applications to restoration of frescoes. Langmuir 17:4251–4255

Angelova L, Terech T, Natali I et al (2011) Cosolvent gel-like materials from partially hydrolyzed poly(vinyl acetate)s and borax. Langmuir 27:11671–11682

Baglioni P, Chelazzi D (eds) (2013) Nanoscience for the conservation of works of art. The Royal Society of Chemistry, Cambridge

Baglioni P, Giorgi R (2006) Soft and hard nanomaterials for restoration and conservation of cultural heritage. Soft Matter 2:293–303

Baglioni M, Giorgi R, Berti D et al (2011) Smart cleaning of cultural heritage: a new challenge for soft nanoscience. Nanoscale 4(1):42–53

Baglioni M, Berti D, Teixeira J et al (2012) Nanostructured surfactant-based systems for the removal of polymers from wall paintings: a small-angle neutron scattering study. Langmuir 28:15193–15202

Baglioni P, Chelazzi D, Giorgi R et al (2013) Colloid and materials science for the conservation of cultural heritage: cleaning, consolidation, and deacidification. Langmuir 29:5110–5122

Bianchin S, Favaro M, Vigato PA et al (2009) The scientific approach to the restoration and monitoring of mural paintings at S. Girolamo Chapel—SS. Annunziata Church in Florence. J Cult Herit 10:379–387

Borgioli L, Caminati G, Gabrielli G et al (1995) Removal of hydrophobic impurities from pictorial surfaces by means of heterogeneous systems. Sci Technol Cult Herit 4(2):67–74

Burnstock A, White R (2000) A preliminary assessment of the aging/degradation of Ethomeen C-12 residues from solvent gel formulations and their potential for inducing changes in resinous paint media. In: Tradition and innovation: advances in conservation, contributions to the Melbourne Congress, 10–14 Oct 2000 (IIC), p 38

Campani E, Casoli A, Cremonesi P et al (2007) L'uso di agarosio e Agar per la preparazione di "Gel Rigidi"-Use of Agarose and Agar for preparing "Rigid Gels. Quaderni del Cesmar7, number 4, Il Prato, Padua

Cappitelli F, Zanardini E, Ranalli G et al (2006) Improved methodology for bioremoval of black crusts on historical stone artworks by use of sulphate-reducing bacteria. Appl Environ Microbiol 72(5):3733–3737

Carretti E, Dei L (2004) Physicochemical characterization of acrylic polymeric resins coating porous materials of artistic interest. Progr Organ Coat 49:282–289

Carretti E, Dei L, Baglioni P (2003) Solubilization of acrylic and vynil polymers in nanocontainer solutions. Application of microemulsions and micelles to Cultural Heritage conservation. Langmuir 19:7867–7872

Carretti E, Giorgi R, Berti D et al (2007) Oil-in-water nanocontainers as low environmental impact cleaning tools for works of art: two case studies. Langmuir 23:6396–6403

Carretti E, Grassi S, Cossalter M et al (2009) Poly(vinyl alcohol)-borate hydro/cosolvent gels: viscoelastic properties, solubilizing power, and application to art conservation. Langmuir 25(15):8656–8662

Carretti E, Natali I, Matarrese C et al (2010) A new family of high viscosity polymeric dispersions for cleaning easel paintings. J Cult Herit 11:373–380

Chelazzi D, Poggi G, Jaidar Y et al (2013) Hydroxide nanoparticles for cultural heritage: consolidation and protection of wall paintings and carbonate materials. J Colloid Interf Sci 392:42–49

Domingues J, Bonelli N, Giorgi R et al (2013) Innovative hydrogels based on semi-interpenetrating p(HEMA)/PVP networks for the cleaning of water-sensitive Cultural Heritage artifacts. Langmuir 29:2746–2755

Domingues J, Bonelli N, Giorgi R et al (2014) Chemical semi-IPN hydrogels for the removal of adhesives from canvas paintings. Appl Phys A 114:705–710

Fantacci S, Amat A, Sgamellotti A (2010) Computational chemistry meets cultural heritage: challenges and perspectives. Acc Chem Res 43(6):802–813

Giorgi R, Dei L, Ceccato M et al (2002) Nanotechnologies for conservation of cultural heritage: paper and canvas deacidification. Langmuir 18:8198–8203

Giorgi R, Bozzi C, Dei L et al (2005) Nanoparticles of $Mg(OH)_2$: synthesis and application to paper conservation. Langmuir 21:8495–8501

Giorgi R, Ambrosi M, Toccafondi N et al (2010a) Nanoparticles for cultural heritage conservation: calcium and barium hydroxide nanoparticles for wall painting consolidation. Chem Eur J 16:9374–9382

Giorgi R, Baglioni M, Berti D et al (2010b) New methodologies for the conservation of cultural heritage: micellar solutions, microemulsions, and hydroxide nanoparticles. Acc Chem Res 43(6):695–704

Gorel F (2010) Assessment of agar gel loaded with micro-emulsion for the cleaning of porous surfaces. CeROArt: Conservation, Exposition, Restauration d'Objets d'Art. http://ceroart.revues.org/1827. Accessed 28 Oct 2013

Iannuccelli S, Sotgiu S (2010) Wet treatments of works of art on paper with rigid Gellan gels. book pap group ann 29, 25–39

Manoudis P, Papadopoulou S, Karapanagiotis L et al (2007) Polymer-silica nanoparticles composite films as protective coatings for stone-based monuments. JPCS 61:1361–1365

Natali I, Carretti E, Angelova L et al (2011) Structural and mechanical properties of "peelable" organoaqueous dispersions with partially hydrolyzed poly(vinyil acetate)-borate networks: applications to cleaning painted surfaces. Langmuir 27(21):13226–13235

Nevin A, Pouli P, Georgiou S et al (2007) Laser conservation of art. Nat Mater 6:320–322

Pizzorusso G, Fratini E, Eiblmeier J et al (2012) Physicochemical characterization of acrylamide/bisacrylamide hydrogels and their application for the conservation of easel paintings. Langmuir 28:3952–3961

Poggi G, Giorgi R, Toccafondi N et al (2010) Hydroxyde nanoparticles for deacidification and concomitant inhibition of iron-gall ink corrosion of paper. Langmuir 26(24):19084–19090

Poggi G, Baglioni P, Giorgi R (2011) Alkaline earth hydroxide nanoparticles for the inhibition of metal gall ink corrosion. Restaurator 32:247–273

Pouli P, Selimis A, Savas G et al (2010) Recent studies of laser science in paintings conservation and research. Acc Chem Res 43(6):771–781

Siano S, Salimbeni R (2010) Advances in laser cleaning of artwork and objects of historical interest: the optimized pulse duration approach. Acc Chem Res 43(6):739–750

Stulik D, Miller D, Khanjian H et al (2004) Solvent gels for the cleaning of works of art: the residue question. Getty Publications, Los Angeles

Wolbers R (2000) Cleaning painted surfaces. Antique Collectors Club, Ltd.

Wolbers R, Sterman N, Stavroudis C (1988) Notes for the workshop on new methods in the cleaning of paintings. The Getty Conservation Institute Marina del Rey, CA. Ottawa: Canadian Conservation Institute, Marina del Rey, CA.

Chapter 2
Consolidation of Wall Paintings and Stone

Abstract Several degradation processes affect immovable works of art such as wall paintings and stone. For example, the loss of binder and the crystallization of salts within the pores of the substrate lead to the detachment of surface layers and the flaking of pigments. Synthetic polymers have been used for consolidation purposes since the 1960s due to their short term appealing performances and a misconception about their stability. However, these materials can undergo degradation, besides they alter the physico–chemical properties of inorganic porous matrices, possibly damaging the works of art. Alternatively, inorganic compounds such as alkaline earth hydroxides can be used for the durable consolidation of artistic and historical objects. The effectiveness of the treatment is highly increased when the size of the consolidants is reduced to the nanoscale, and dispersions of nanoparticles in solvents (e.g. alcohols) exhibit good penetration and consolidating power. This chapter illustrates the main formulations of nanomaterials that have been developed in the last decades for the consolidation of wall paintings and stone, providing references to real case studies and application fields. Finally, a whole section is dedicated to applicative procedures and protocols for the use of nanoparticles dispersions.

2.1 Background

This chapter deals with the use of nanoparticles dispersions for the consolidation of immovable works of art such as wall paintings and stone, discussing the advantages and the practical aspects of this methodology. The readership of this book will be probably accustomed with the structure, the main composition and the properties of both wall paintings and stone used in historical and artistic heritage. Therefore, we will briefly recall only the fundamental notions that serve as a basis for illustrating the application of nanomaterials.

Typically, a wall painting belonging to the classic tradition consists of three different superimposed layers (see Fig. 2.1). The most inner layer, called *arriccio*, is laid directly on the stone or brick wall, and usually consists of a 3:1 (v/v) mixture of coarse sand and lime paste ($Ca(OH)_2$ and water). The surface roughness of the

P. Baglioni et al., *Nanotechnologies in the Conservation of Cultural Heritage*,
DOI 10.1007/978-94-017-9303-2_2

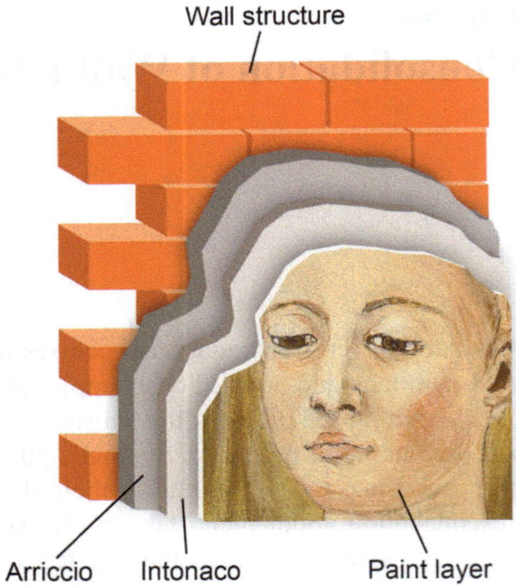

Fig. 2.1 Stratigraphy of a model wall painting belonging to the classic tradition. The most inner layer (*arriccio*) is laid directly on the stone or brick wall. The *intonaco* is applied over the *arriccio* obtaining a smooth surface over which the artist paints. The painted layer is the most external layer, whose thickness is of ca. 50–500 μm. (Image by Michele Baglioni)

arriccio (Italian word for "curl"), and consequently its surface area, is enhanced so to improve the adhesion of the second layer, which is called *intonaco* (Italian for "lime plaster"). When the *arriccio* is completely dry, the *intonaco* is prepared using a 2:1 or 1:1 (v/v) mixture of finer sand particles (micrometer size) and lime, and then applied obtaining a very smooth surface over which the artist paints using pigments. The painted layer is indeed the third and most external layer, whose thickness is of ca. 50–500 μm. In the so called *a fresco* technique, the pigments are dispersed in lime water (a saturated water solution of $Ca(OH)_2$) and applied directly on the wet *intonaco* before it sets, in fact "*a fresco*" is the Italian expression for "in wet conditions". The setting process, also called "carbonation", is the reaction of $Ca(OH)_2$ with atmospheric carbon dioxide (CO_2), to form a solid network of calcium carbonate ($CaCO_3$) that acts as a binder and embeds the pigments granting optimal mechanical strength to the painted layer:

$$Ca(OH)_2 + CO_2 \rightarrow CaCO_3 + H_2O$$

However, the pH of the wet *intonaco* is highly alkaline, and some pigments are not compatible with such alkalinity, therefore they are applied after the completion of the carbonation process, i.e. *a secco* (in dry conditions). In this case the binder of pigments is not $CaCO_3$, but a material of either vegetal (gums, extractives, oil) or animal origin (casein, egg, animal glue). A third technique is the so-called *mezzo fresco* ("half wet"), where the artist paints on nearly dry *intonaco*. Artists also used mixed *a fresco* and *a secco* techniques.

Carbonate-based wall paintings, independently of the technique used, were diffused in the ancient Northern Africa and Near East (Morocco, Egypt, Crete) and in Europe during classical antiquity (ancient Greece, Rome), the Middle Ages, the

Renaissance and early modern age. However, their presence is not restricted to Europe and the Mediterranean basin, but extends also to Asia (e.g. India, China) and Southern-America. Nonetheless, there are also cases where the painted layer exhibits a high silicate (clay) content, rather than carbonate, as for instance in several pre-Hispanic mural paintings from central Mexico. Clays were probably used to polish the painting surface, or as a support for dyes.

Regarding stone, carbonate-based rocks such as limestone and marble have been used throughout the ages as building and artistic materials. Depending on the areas, sandy limestone or sandstone has also been used. The chemical composition and the physico–chemical properties of stone can vary greatly according to its origin, location and use, and the full description of the various types of stone is beyond the scopes of this Compendium.

Wall paintings and stone undergo several degradation processes that include physical erosion, freeze-thaw cycles, biodegradation and chemical corrosion by acid rain and moisture. For instance, pollution gases in the atmosphere, such as sulfur and nitrogen oxides, can hydrate and form acids. One of the most diffused and critical processes is the formation of aqueous sulfuric acid (H_2SO_4) upon hydration of SO_2, which is released in the combustion of sulfide-containing materials and mineral oils. Sulfuric acid corrodes calcium carbonate to form gypsum, according to the following reaction:

$$H_2SO_4 + CaCO_3 + H_2O \rightarrow CaSO_4 \cdot 2H_2O + CO_2$$

It is important to note that the crystalline unit cell of gypsum ($CaSO_4 \cdot 2H_2O$) is bigger than that of calcium carbonate; therefore the formation of gypsum crystals produces mechanical stress within the pores of the work of art. Moreover, gypsum is a fairly soluble salt, thus it can spread as a water solution through porous matrices and undergo crystallization–dissolution cycles owing to hydrothermal oscillations. The presence of salts can also be due to water solutions rising from the ground and diffusing through the pores of the work of art. Besides sulfates, other salts such as alkaline and alkaline earth chlorides, nitrates, and oxalates can be present in the pores of wall paintings and stone, leading to harmful effects (Arnold 1987; Piqué et al. 1992). For instance, tetrahydrate calcium nitrate ($Ca(NO_3)_2 \cdot 4H_2O$) is commonly found in wall paintings, and being a deliquescent salt it can produce water condensation on the substrate from moist air at temperatures well above the saturation value.

The corrosion of calcium carbonate and the mechanical stress due to salts crystallization jeopardize the surface of wall paintings, which can exhibit efflorescence (salt crystal blooms) and flaking painted layers, as shown in Fig. 2.2. Throughout the ages, a variety of different materials has been applied onto wall paintings, as consolidants, glues, adhesives and fixatives. Starting from the 1960s, synthetic products such as acrylate, vinyl, silicone and epoxy polymers have been widely used based on appealing short term performances and a misconception about their stability. The high adhesive power of synthetic formulations made them a fast, easy and suitable choice for the re-adhesion of flaking parts. Moreover, the hydrophobic

Fig. 2.2 Salt efflorescences on the surface of wall paintings in the site of Ixcaquixtla (Mexico). The picture shows the corrosion of calcium carbonate and the mechanical stress due to salts crystallization. (Reprinted from Chelazzi et al. 2013. Copyright 2013, with permission from Elsevier.)

character of these products justified their use as protective coatings meant to prevent degradation caused by water-related chemical and physical processes, and the coatings also provided saturation of colors with appealing aesthetic effects. Finally, owing to the higher resistance of synthetic polymers to aging as compared to natural resins, conservators were led to think that the application of synthetic products would be fully reversible even in the long term.

Unfortunately, practical evidence over the past 50 years has shown that this is not the case.

As a matter of fact, synthetic polymers can undergo degradation due to the action of temperature, relative humidity, saline solutions and UV-Vis light, which produce aesthetical alteration (discoloration) and change in the polymers solubility. Cross-linking and chain scissions can take place so that the polymers molecular weight is altered and consequently their solubility in the original solvents is decreased (Lazzari and Chiantore 2000; Chiantore and Lazzari 2001; Favaro et al. 2006). In some cases, the alteration of the coatings can be such that their removal from the artistic/historical surfaces is intractable with traditional solvents, and alternative technology must be used, for instance nanostructured cleaning fluids (Baglioni et al. 2012). The removal of aged and detrimental coatings from wall paintings and stone using nanotechnology will be described in Chap. 3.

Besides the possibility of undergoing degradation themselves, polymer coatings strongly alter the physico–chemical properties of inorganic porous substrates. A study by Carretti and Dei (2004) investigated the modifications produced by films of Paraloid B72®, Primal AC33® and Elvacite 2046® on the surface of aerial mortar samples simulating real wall paintings. The main effects were the drastic decrease (40–50 %) of the surface water vapor permeability and hydrophilicity, meaning that water transpiration at the interface between the porous samples and the external

Fig. 2.3 Mural paintings treated with Paraloid B72® (archaeological site of Monte Alban, Mexico). The pictures highlight aesthetic alteration of the surface (gloss), disruption and detachment of the painted layer. (Photographs by Rodorico Giorgi, printed with the kind permission of INAH-CNCPC.)

environment was strongly hampered, independently on the copolymer applied. The alteration of the surface "breathing" is highly detrimental: the permanence time of water and salt solutions within the porous matrix increases, leading to larger salt crystals that form in the pores and produce mechanical stress beneath the painted layer. As a result, the pictorial surface of the work of art can be disrupted (see Fig. 2.3) in a period of time ranging roughly from less than 10 to 50 years, the degradation being enhanced by fluctuations of temperature and relative humidity, and by pollution (Giorgi et al. 2010a). For instance, the combination of the presence of a Mowilith film (vinyl acetate/acrylate copolymer) and salt crystallization produced in less than 10 years the irreversible loss of portions of original wall paintings in the archaeological site of Mayapan (Mexico, Post-Classic Age 1200–1450 BC), and degradation of paintings (e.g. flaking) has been reported for several other Mexican sites where different polymers had been applied as fixatives in past restoration interventions (Espinosa 1987; Riederer 1984; López and Dávila 2002). European works of art can exhibit similar degradation processes after treatment with polymers (Carretti et al. 2007; Grassi et al. 2007), even though the process normally occurs at a slower rate because the environmental conditions are less extreme.

Therefore, synthetic adhesives should be used only when strictly necessary, and alternative methods have been proposed in the last decades to restore wall paintings and stone, focusing on the use of materials whose physico-chemical properties are similar to (or possibly the same as) those of the work of art, i.e. "compatible" materials.

The physical chemist Enzo Ferroni pioneered the use of compatible materials at the end of the 1960s, when he developed several solutions to the conservation issues that threatened part of the Florence patrimony following the tragic events of the 1966 Florence flood. Ferroni realized a method, named after him, for the removal of sulfates and the consolidation of wall paintings based on the application in two successive steps of ammonium carbonate and barium hydroxide aqueous solutions loaded in cellulose poultices (Ferroni et al. 1969; Matteini and Moles 1984; Ferroni and Baglioni 1986; Matteini 1991; Baglioni and Giorgi 2013). The Ferroni method will be described in details in Sect. 2.4 (Application procedures) since it is the precursor of the development and use of calcium hydroxide nanoparticles for the consolidation of wall paintings, and it should be considered as a preliminary before the application of nanoparticles on paintings that are heavily polluted by sulfates. The application of the method leads to two fundamental results that can be resumed as follows: (a) the extraction of gypsum and the transformation of gypsum residues into stable and insoluble barium sulfate; (b) the regeneration of calcium hydroxide ($Ca(OH)_2$) within the pores of the painted layer; $Ca(OH)_2$ then reacts with CO_2 forming calcium carbonate that provides consolidation without altering the chemical composition of the artifact. The Ferroni method was applied for the first time on the Beato Angelico wall paintings in Florence. The conservator Dino Dini carried out the paintings restoration, and his partnership with Ferroni stands as an example of interdisciplinary cooperation for the preservation of Cultural Heritage. Since then, the method has been successfully applied worldwide, and has proven effective even on artifacts exposed to extreme hydrothermal conditions. For instance, it has been used for the consolidation of mural paintings in the Mexican site of Cacaxtla, and the restored works of art are still in good conditions while paintings in the same site treated about 20 years ago with Paraloid® show nowadays severe degradation (Baglioni and Giorgi 2013).

In fact, calcium hydroxide is the most compatible material for the consolidation of carbonate-based works of art. However, the solubility of $Ca(OH)_2$ in water is scarce (1.7 g/L at 20 °C), therefore even a saturated calcium hydroxide aqueous solution (limewater) has an extremely poor consolidating power, and introducing large amounts of water in wall paintings or stone is not advisable because it would encourage degradation processes such as freeze-thaw cycles, solubilization/crystallization of salts and biodegradation. On the other hand if dispersions of solid particles are considered, rather than ionic solutions, it is possible to increase the "concentration" of calcium hydroxide. The fundamental idea is then to apply the dispersion of solid particles over the artistic/historical substrate (using a brush, syringe, sprayer, etc.); the hydroxide particles penetrate within the substrate pores and re-create a network of crystalline $CaCO_3$ upon carbonation, bridging flaking parts. Therefore, the newly formed calcium carbonate is not merely a filler, but a true consolidant.

Aqueous dispersions of $Ca(OH)_2$ particles are not stable, as the hydroxide particles undergo fast aggregation through bridging driven by hydrogen bonds (Gregory

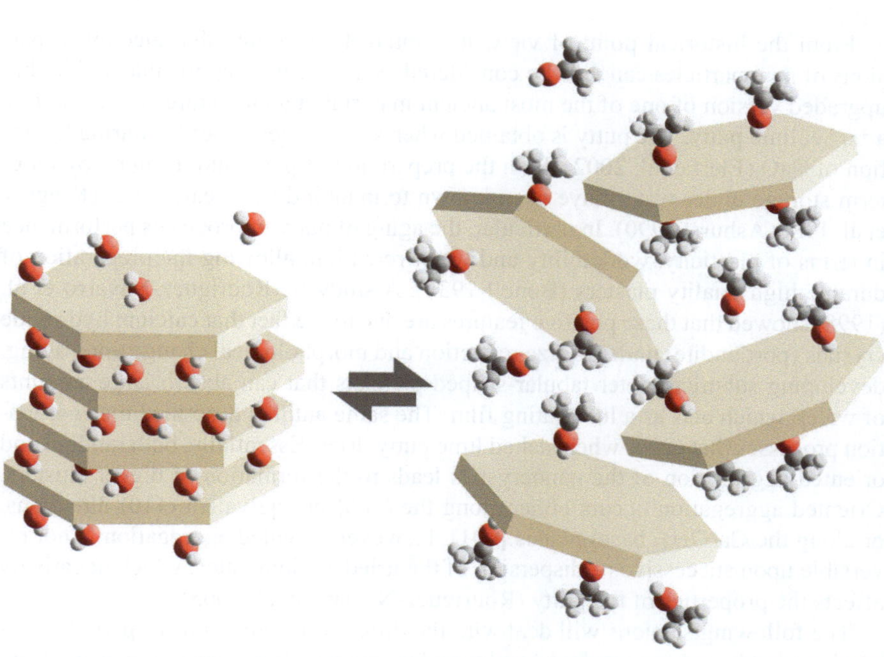

Fig. 2.4 (*Left*) Aggregation of calcium hydroxide nanoparticles in water through bridging driven by hydrogen bonds. (*Right*) The particles face-to-face sticking is inhibited due to the absorption of 2-propanol, leading to stabilization of the dispersion. (Image by Michele Baglioni)

and Tadros 1987), see also Fig. 2.4. This results in sedimentation, which prevents penetration through porous matrices, leading to scarce consolidation and to the formation of white hazes on the treated surface. Therefore the main use of aqueous dispersions of lime has been limited to internal injections or micro-stuccoing (Peroni et al. 1982). The stability of lime particles in water might be increased using surfactants, but this would involve leaving surfactant residues in the porous matrix, which should be avoided, as much as possible.

On the other hand, the use of short chain alcohols as dispersion media increases stability. Starting from the beginning of the 1990s, researchers at CSGI (Consorzio Interuniversitario per lo Sviluppo dei Sistemi a Grande Interfase; Research Center for Colloid and Surface Science, established in Florence by Ferroni in 1993) developed the first formulation of calcium hydroxide particles in propanol, and the use of stable $Ca(OH)_2$ dispersions for Cultural Heritage preservation was later patented (Baglioni et al. 1996). Since then, numerous refinements and improvements have been carried out, and a commercial formulation of calcium hydroxide nanoparticles in 2-propanol is available on the market since 2008 under the trademark Nanorestore® (from: CTS Srl., Altavilla Vicentina/Italy. www.ctseurope.com). Starting from this background, other research groups have also developed and applied nanomaterials for the consolidation of immovable works of art (see next sections).

From the historical point of view, it is interesting to note that alcohol dispersions of nanoparticles can also be considered as a new, fully applicable and highly upgraded version of one of the most ancient material in architecture and decorative arts, i.e. lime putty. The putty is obtained when water excess is added during hydration of CaO (Elert et al. 2002); both the preparation of putty and its aging by long-term storage under water have been known to mankind since early ages (Kingery et al. 1988; Ashurst 1990). In particular, the aging of putty improves its performance in terms of plasticity, workability and water retention, allowing the preparation of durable high quality plasters (Bonell 1934). A study by Rodriguez-Navarro et al. (1998) showed that these positive features are due to the fact that calcium hydroxide crystals (portlandite) undergo size reduction and morphological change upon aging, developing submicrometer tabular-shaped platelets that can absorb large amounts of water, which acts as a lubricating film. The same authors described the aggregation processes that occur when slaked lime putty dries. Essentially, both random and oriented aggregation of the nanocrystals leads to the formation of bigger clusters. Oriented aggregation occurs either along the ⟨100⟩ or equivalent ⟨110⟩ directions, or along the $Ca(OH)_2$ basal planes [001]. However, oriented aggregation is not reversible upon successive re-dispersion of the dried agglomerates, which negatively affects the properties of the putty (Rodriguez-Navarro et al. 2005).

The following sections will deal with the different formulations of particles dispersions that have been produced and used for consolidation purposes, starting from fundamental research work. The main experimental and practical results will be considered. The influence of environmental conditions will also be discussed in order to define the possible application fields. Finally, practical guidelines will be provided, based on the experience acquired in the last decades, commenting on applicative aspects and possible issues.

2.2 Formulations

The scientific literature reports numerous studies on the synthesis of metal oxides and hydroxides nanoparticles, and a thorough description of this vast topic is beyond the scopes of this Compendium. Therefore, this section will focus exclusively on the formulations of nanoparticles dispersions that have been proposed and applied for the consolidation of wall paintings and stone throughout the last decades. Besides describing the main formulations that are already on the market, we decided to report also some of the fundamental research work from which the commercial products have stemmed. Indeed, we believe that it is important to track the intellectual path that has led to the development of technological tools for the preservation of Cultural Heritage. Therefore, both systems that have been produced so far on a laboratory or semi-industrial scale and market products will be considered, mentioning applicative case studies, so to provide the reader with a complete scenario.

2.2.1 *Formulations for Wall Paintings*

As mentioned in the previous section, research on compatible inorganic materials such as micro- and nano-particles for the consolidation of wall paintings started at CSGI, in Florence, at the beginning of the 1990s. The first application reported in the literature by Giorgi et al. (2000) shown that slaked lime particles with an average size of 3–4 μm could be dispersed by vigorous stirring in 1-propanol, and the obtained dispersion had a stability of many hours, allowing application to wall paintings. The calcium hydroxide concentration was more than three times that of limewater. The authors tested the dispersion on model mortar samples that simulated severe decohesion (low binder concentration). The application resulted in high re-cohesion of the upper painted layers, as showed by resistance to tangential scratches and surface abrasion test. The dispersion was then tested on a portion of a wall painting belonging to the Santa Maria Novella church in Florence. The painted layer was powdering and flaking. Two applications of the dispersion (5 g/L), each followed by 7–10 days of setting, provided the strengthening of the painted layer that was then treated with the Ferroni method, with excellent overall effects (Fig. 2.5). It is important to notice that the whole restoration process was carried out without introducing any organic material, so to preserve the physico–chemical properties of the artistic substrate. Ambrosi et al. (2001b) further investigated the kinetic stability in alcohols of $Ca(OH)_2$ micro-particles (1–2 μm) obtained as described above, and reported that the stability enhancement is in the order 1-propanol > ethanol > 2-propanol, which suggested that the stability is proportional to the thickness of the hydrophobic layer on the hydroxide particles (that adsorb alcohol molecules). In fact, it is known that stability is affected by the adsorption of solvent molecules on the particles at the solid/liquid interface (Vincent et al. 1971). This aspect will be further discussed below.

Besides allowing the stable dispersion of nanoparticles, good solvents for the treatment of wall paintings and stone should exhibit a not too high volatility; otherwise particles penetration would be hampered. On the other hand, low-volatility solvents with high surface tension and boiling point would remain in the substrate pores, slowing excessively the consolidation process. Low viscosity and surface tension values allow better wetting of matrices. Short chain alcohols such as ethanol and propanol represent a good balance for practical applications.

The natural upgrade of the aforementioned applications was the synthesis of nano-sized $Ca(OH)_2$ particles and their stable dispersion in alcohols. The main rationale is that reducing the particles size—from microns to nanometers—grants higher stability of the dispersions and increases the penetration of the particles through porous matrices (hence it reduces the risk of forming white hazes on the surface). Moreover, smaller particles exhibit a higher reactivity to form a $CaCO_3$ crystalline network that grants cohesion and mechanical strength to the substrate layers, embedding and fixing flaking grains and pigments.

Different methods have been proposed for the preparation of $Ca(OH)_2$ nanoparticles dispersions in alcohols, and numerous applications of these systems have been

Fig. 2.5 Portion of wall paintings by Andrea da Firenze (fourteenth century), in the Spanish Chapel of the green Cloister (*Chiostro Verde*) of the Santa Maria Novella church (Florence, Italy). (*Top*) the powdering and flaking painted layer before restoration. (*Bottom*) the right half of the wall painting portion after pre-consolidation with a dispersion of $Ca(OH)_2$ micro-particles in 1-propanol. (Reprinted from Giorgi et al. (2000), Fig. 5. http://www.maneyonline.com/doi/abs/10.1179/sic.2000.45.3.154, with permission from International Institute for Conservation of Historic and Artistic Works (IIC) and Maney Publishing)

carried out in the framework of wall paintings and stone conservation, as also reported by reviews in the literature (see for instance reviews by Chelazzi et al. 2013; Baglioni and Giorgi 2006). Besides particles size, several other characteristics influence the behavior of nanoparticles in consolidating carbonate-based porous matrices. Examples include:

- The particles size and size distribution (dispersity)
- The solvent used for dispersing the particles, and its interaction with both the particles surface (e.g. adsorption) and the artistic substrate (wetting properties)
- The particles crystal habitus and crystallinity
- The crystalline domains size and the presence of defects
- The particles surface area

Research has focused on the synthetic pathways, the structural characterization of particles, and on the stability of the dispersions. Moreover, the interaction of the particles with the artistic and historical substrates has been investigated, based on the practical evidence of the effectiveness of these systems. Nonetheless, there are several aspects that need further investigation in order to grant the full control of the applications, and each real case study exhibits its own difficulties and characteristics.

The fundamental processes for the synthesis of nanoparticles can be roughly divided into the so called "bottom–up" and "top–down" approaches. The first method consists in having the nanoparticles built up by the collection of atoms or ions, or by the deposition and growth of crystals for instance from liquid phase. On the other hand, in the top–down approach bigger particles are broken down to the nanoscale, e.g. by grinding or milling.

Ambrosi et al. (2001a) reported the bottom–up synthesis of calcium hydroxide nanoparticles from aqueous homogeneous phase, and their application to *frescoes* restoration. The particles were precipitated by rapidly mixing aqueous solutions of sodium hydroxide (NaOH) and calcium chloride ($CaCl_2$) at 90 °C, under stirring, at high supersaturation degree (S), which is defined as:

$$S = \left[Ca^{2+}\right] / \left[Ca^{2+}\right]_{sat}$$

where $[Ca^{2+}]_{sat}$ is the concentration of Ca^{2+} ions in the $Ca(OH)_2$ saturated solution. The aqueous suspension of hydroxide particles is allowed to reach room temperature under a nitrogen atmosphere, to prevent carbonation. The supernatant is then discarded and the remaining suspension is washed several times with limewater to reduce the concentration of the byproduct NaCl. The suspension is then concentrated under vacuum at 40 °C. Finally, the particles are dispersed in 1-propanol (*n*-propanol). The particles obtained by this method are crystalline (portlandite) hexagonal platelets, with hexagon sides of 100–300 nm and thickness of 2–40 nm (see Fig. 2.6). The dispersion of particles in 1-propanol exhibits good kinetic stability over periods of 20 h, which allowed the application to a degraded sixteenth century wall painting by Santi di Tito, in the Cathedral of Florence ("Il Duomo", i.e. the dome). The nanoparticles are stable for longer time with respect to commercial micron-sized calcium hydroxide, and the small size of the particles can penetrate in the painted layer, avoiding the formation of white hazes on the painting surface. Moreover, the hexagonal nano-platelets have a great capacity to absorb water, which favors the carbonation process. In fact, the application of nanoparticles granted the consolidation of the flaking and powdering painted layer, as shown in Fig. 2.6.

Alkaline nanoparticles for the deacidification of works of art were patented by Baglioni et al. in 2002. Between 2002 and 2005, Giorgi et al. (2002; 2005a, b) reported the use of calcium and magnesium hydroxide ($Mg(OH)_2$) nanoparticles obtained from aqueous homogeneous phase synthesis and dispersed in propanol for the deacidification of paper, canvas and wood, which will be detailed in Sect. 5.2. It is important to notice that mixed dispersions of $Mg(OH)_2$ and $Ca(OH)_2$ nanoparticles can also be used for the consolidation of dolomite stone (see Sect. 2.2.2).

Salvadori and Dei (2001) investigated a different synthetic route to decrease the $Ca(OH)_2$ particles size, based on the fact that in synthetic processes temperatures above 100 °C (in non-aqueous media) are known to promote the formation of nano-sized particles (Yura et al. 1990) and that organic solvents affect the size and shape of the precipitated particles (Hamada and Matijevic 1982; Matijevic and Cimas 1987). Therefore, hydrate calcium chloride ($CaCl_2{\cdot}2H_2O$) was solubilized in ethanediol (ED) or propanediol (PD) at high temperature (either 150 or 175 °C),

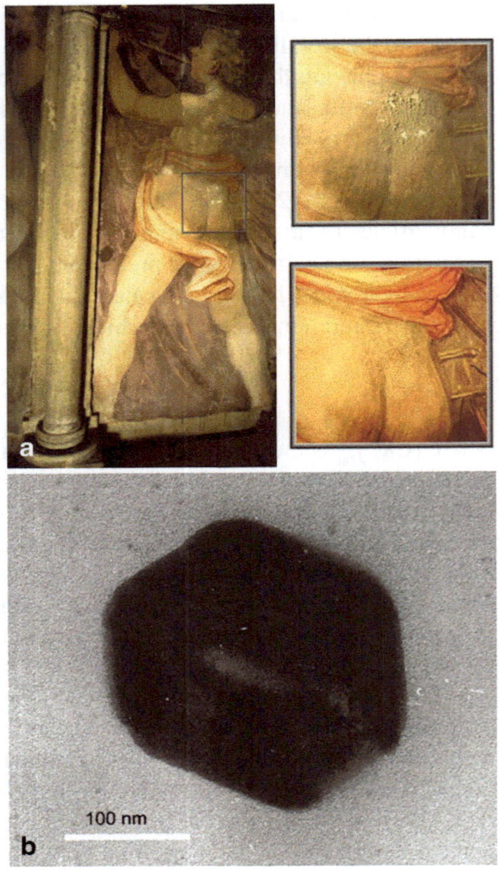

Fig. 2.6 **a** Wall paintings by Santi di Tito (sixteenth century) on the counterfaçade of the Florence Cathedral ("Duomo"). The region labeled with the box was treated with $Ca(OH)_2$ nanoparticles. (*Top box*) before restoration. (*Bottom box*) after restoration. **b** SEM image of a $Ca(OH)_2$ nanoparticle with hexagonal habitus, obtained from homogeneous phase reaction (*bottom–up*). (Reprinted with permission from Ambrosi et al 2001a. Copyright 2001 American Chemical Society.)

and then an aqueous NaOH solution was added dropwise to precipitate the calcium hydroxide particles. The system was aged under stirring for some minutes. Finally, the particles were separated from the surnatant and dispersed with 2-propanol in an ultrasonic bath, to remove the diols that were adsorbed onto the particles. Several reactions parameters were varied, monitoring the changes in the dimension of the obtained particles. In particular, short aging time produced very small and almost spherical particles (30–60 nm); otherwise the syntheses produced hexagonal platelets whose size ranged from 50 to 150 nm depending on the molar ratio of the reactants (see Fig. 2.7). XRD measurements showed that the synthesized particles are crystalline and suggested that 2-propanol is adsorbed on the basal face of the hexagonal platelets. In particular, it is the hydroxyl group (–OH) of the alcohol molecules that adsorbs (physisorption) to the hydroxide particles surface (Salvadori and Dei 2001; Fratini et al. 2007). According to Fratini et al. (2007), "[…] the specific surface is governed by the hydrophobicity of the surface layer that in the presence of 2-propanol favors smaller particles by slowing down growth on the

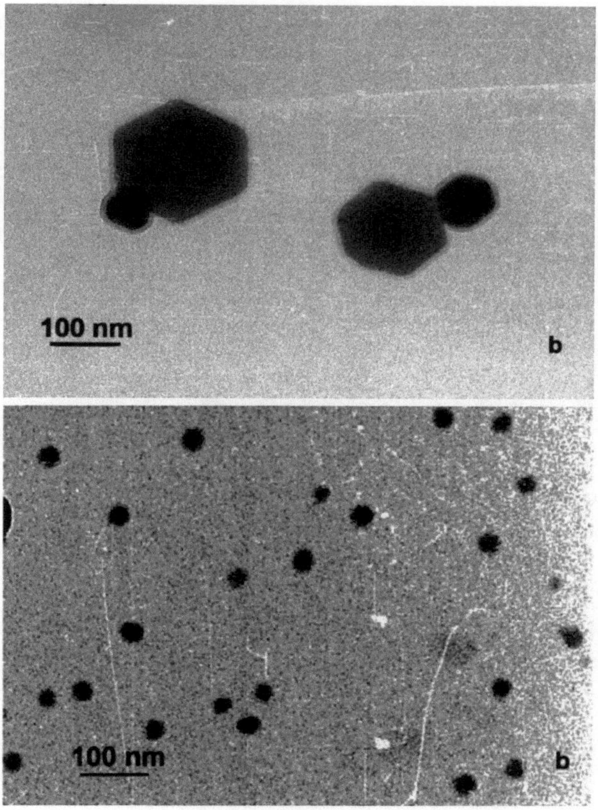

Fig. 2.7 TEM images of hexagonal (*top*) and spherical (*bottom*) $Ca(OH)_2$ nanoparticles obtained through homogeneous phase reactions (*bottom–up*) in diols. (Reprinted with permission from Salvadori and Dei 2001. Copyright 2001 American Chemical Society)

surface relative to nucleation of separate particles". Moreover, the particles face-to-face sticking is inhibited due to absorption of 2-propanol. Overall the addition of 2-propanol leads to the stabilization of nanoparticles (see Fig. 2.4). It is interesting to notice that upon drying, the particles tend to pile along the basal face (the {001} plane) as evidenced by XRD. In this way, particles form layers within the porous matrix of the substrate, and then undergo carbonation forming the $CaCO_3$ network that provides consolidation.

Later, Nanni and Dei (2003) carried out the synthesis of $Ca(OH)_2$ nanoparticles in water-in-oil (w/o) microemulsions where the oil phase was cyclohexane. Two w/o emulsions are mixed, containing respectively Ca^{2+} and OH^- ions in the dispersed water droplets. The nano-sized water droplets dispersed in cyclohexane serve thus as templates: the formation of solid nanoparticles takes place within the droplets, and as a result the final particle size is very small (2–10 nm, see Fig. 2.8). The most critical factor is the choice of the suitable microemulsion composition. However, XRD measurements showed that particles with such a small size are highly reactive to atmospheric CO_2. In fact, for such small particles it is hard to distinguish between surface and bulk carbonation, and even the mere manipulation required for the XRD

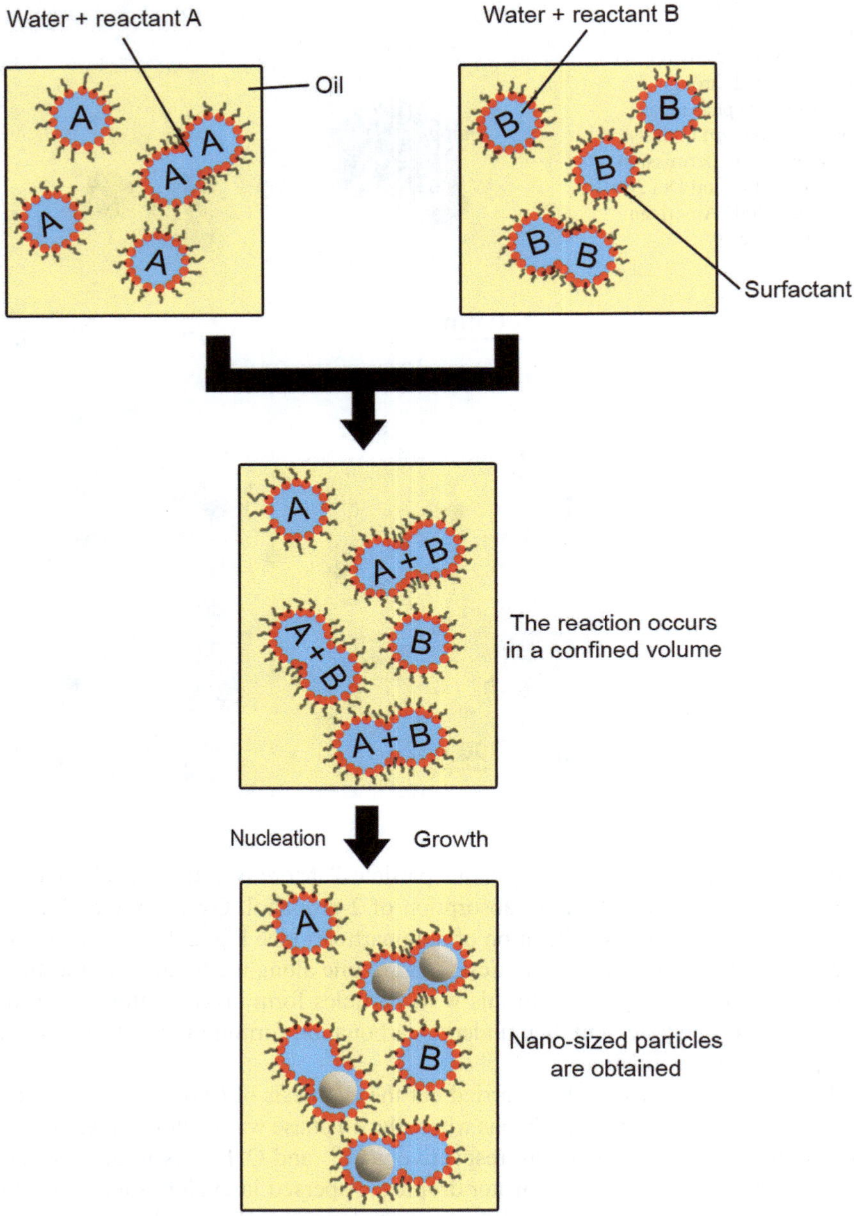

Fig. 2.8 Synthesis of $Ca(OH)_2$ nanoparticles in water-in-oil (w/o) microemulsions. Two w/o emulsions are mixed, containing respectively Ca^{2+} **(A)** and OH^- **(B)** ions in the dispersed water droplets, which serve as templates: the formation of solid nanoparticles takes place within the droplets. (Image by Michele Baglioni)

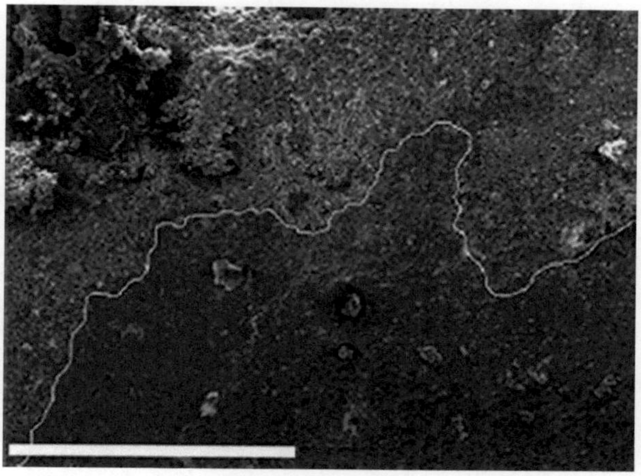

Fig. 2.9 SEM image of a degraded fresco specimen untreated (*above the white line*) and treated with $Ca(OH)_2$ nanoparticles dispersion in 2-propanol (*under the white line*). The treatment resulted in renewed cohesion of the powdering surface, which appeared smoother. Bar is 400 μm. (Reprinted with permission from Dei and Salvadori 2006. Copyright © 2006 Elsevier Masson SAS. All rights reserved)

analysis resulted in the complete transformation of $Ca(OH)_2$ in $CaCO_3$. Therefore, the use of these particles for the consolidation of wall paintings could be hampered by a too fast reactivity, and this synthesis has been instead considered to prepare ultra-fine calcium carbonate particles.

However, it must be considered that both the synthesis in diols and in w/o microemulsions are time-consuming and the latter method has a low production yield. Nonetheless, the two studies provided fundamental knowledge on the particles behavior.

Dei and Salvadori (2006) further tested dispersions of $Ca(OH)_2$ nanoparticles in 2-propanol for the consolidation of wall paintings. The dispersions were applied on *fresco* specimens mimicking color powdering, using different concentrations (5 g/L and lower). Analysis with scanning electron microscopy (SEM) showed that the treatment resulted in renewed cohesion of the powdering surface, which appeared smoother (see Fig. 2.9). Energy dispersive X-ray Spectroscopy (EDX) confirmed that the treated samples had higher binder content (higher Ca/Si ratio). Based on these results, diluted dispersions (0.05 g/L) were applied on a thirteenth century fresco in the Saint Zeno Church in Verona (Italy). After carbonation of the particles, the consolidation effect was positively assessed, and visual inspection under grazing light highlighted that the treatment had produced cohesion and smoothing of the surface.

The nanoparticle treatment was thus extended to large portions of the paintings, with positive results in terms of both re-cohesive and aesthetical effects. Dei et al. (2005, 2007) applied nanoparticles on wall paintings at the Bargello Palace (Museo Nazionale del Bargello—National Museum of Bargello, Florence, Italy), and provided additional details on applicative aspects in a paper regarding the treatment of a wall painting by Agnolo Gaddi (fourteenth century) in the church of Santa Croce (Basilica of the Holy Cross, Florence, Italy). The application of $Ca(OH)_2$ nanoparticles has been reported also for the consolidation of stucco, where reinforcement of the surface layer was needed (Baglioni et al. 2001; Dei et al. 2001).

Fig. 2.10 **a** Application of $Ca(OH)_2$ nanoparticles dispersion in 2-propanol for the consolidation of graffiti in the prison of the historical Palazzo Chiaramonte-Steri (Palermo, Italy). The particles were applied by brush on a sheet of Japanese paper. **b, c** Photographs at grazing light of mural lunettes at the "SS. Giuda and Simone" cloister (Corniola, Empoli-Florence, Italy); **b** the powdering mural surface before treatment; **c** the surface after treatment with $Ca(OH)_2$ nanoparticles dispersion in 2-propanol. (Readapted with permission from Natali et al. 2014. Copyright © 2014 Elsevier Masson SAS. All rights reserved)

The commercial product Nanorestore® (dispersion of calcium hydroxide nanoparticles in 2-propanol, 5 g/L, nanoparticle size around 250 nm), which stemmed from all the aforementioned research and practical activities, has been tested in numerous real case studies. For instance, Natali et al. (2014) reported the use of this product for the consolidation of historical graffiti in Palazzo Chiaromonte-Steri (Palermo, Italy, see Fig. 2.10), a building begun in the early fourteenth century, and used as a residence by the Sicilian lord Manfredi Chiaromonte III, Count of the powerful feud of Modica. From 1600 to 1782 the building hosted the court of the Inquisition, and the prison's walls were covered by the graffiti of the prisoners and victims, written in ancient Sicilian dialect, Hebrew and English. The graffiti were painted directly on the wall with a red pigment (ground clay brick) dispersed in an organic medium, and were discovered under the plaster in 2005, during the restoration of the palace. The mural surface was heavily jeopardized due to the presence of salt efflorescence and the pigment was powdering. To achieve consolidation of the highly damaged painted layer it was preferred to carry out several treatments at low concentration (0.06% wt/vol), and no whitening effects were observed on the treated surface even after ten successive applications. Analysis through Fourier Transform Infrared Spectroscopy (FTIR) showed that carbonation of the calcium hydroxide particles took place in 8–10 days, and as a consequence the painted surface became more compact and strengthened. EDX analysis showed an increase content in calcium in the plaster surface layers, confirming that penetration of the $Ca(OH)_2$ nanoparticles had occurred, and water absorption tests highlighted that the wall porosity and water exchange capacity had not been altered by the treatment. The same formulation was applied to eighteenth century mural lunettes belonging to the Saint Jude and Simon cloister in Corniola (Empoli—Florence, Italy). The lunettes exhibited lacunas and powdering of the painted layer. The application of nanoparticles resulted in the strengthening of the pictorial surface, as highlighted both by visual assessment (Fig. 2.10) and through the scotch tape test: the powdering color film was completely re-adhered to the substrate and the amount of grains removed by the tape was practically tending to zero after completion of the treatment. It is interesting to note that in this case some applications were required before the consolidation effect became evident, which suggested that nanoparticles initially penetrated deeply within the substrate, and then consolidation took place "from inner to outer".

The Nanorestore® product proved effective also for the consolidation of mural paintings in areas where the environmental conditions (T and RH) are such that degradation of the works of art occurs at a faster rate. For instance, wall paintings in the Mesoamerican archaeological sites of Calakmul (Campeche, Mexico) and Ixcaquixtla (Puebla, Mixtec area, Mexico) were effectively consolidated, fixing the pigments that exhibited severe degradation due to typical processes such as salt crystallization and formation of biofilms (Baglioni and Giorgi 2006; Giorgi et al. 2010b; Baglioni et al. 2013). For the Maya wall paintings in Calakmul a mixed formulation of calcium hydroxide and barium hydroxide nanoparticles (see Fig. 2.11) was also used, to consolidate substrates where large amounts of soluble sulfates were present (Baglioni et al. 2014), see also Sect. 2.3. The calcium hydroxide particles were obtained using a top–down approach: lime putty typically has some CaO residue within the hydroxide particles core; it is then possible to achieve the completion of the particles hydration (slaking) via a thermo-mechanical treatment, under controlled conditions in terms of temperature and pressure. The transformation of CaO into $Ca(OH)_2$ in the core of the particles leads to volume expansion and fragmentation of the particles, which are reduced to the nano-size, i.e. 100–300 nm (the process does not need further steps such as purification etc.). By modifying the process parameters it is possible to obtain also a second minor population of particles (besides the main one) whose size is about 1 μm, which can be useful for the consolidation of highly porous substrates. The top–down approach was used also for the preparation of barium hydroxide nanoparticles (~100 nm, see Fig. 2.11), since in this case the hydroxide solubility in water (3.89 g/100 mL at 20 °C) hinders the bottom–up process.

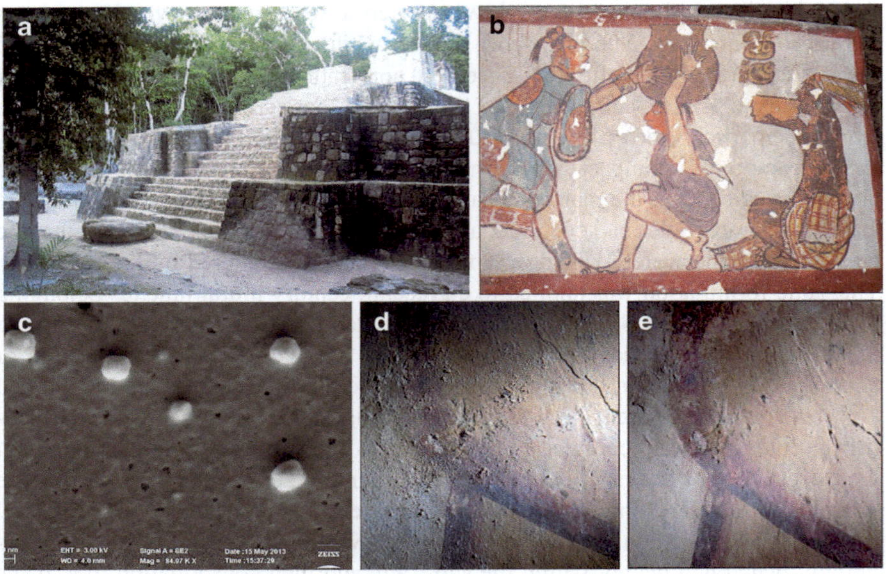

Fig. 2.11 **a** Structure I in the "Acropolis Chik-Naab" in the Calakmul archaeological site (Campeche, Mexico). **b** Mural painting from the Early Classic Maya period, decorating the first step of sub-structure I in the south–east side of the building. **c** $Ba(OH)_2$ nanoparticles obtained through a top–down approach; bar is 200 nm. **d** Details of the Maya mural paintings showing the presence of sulfates over the surface and flaking phenomena damaging the paint layer. **e** The same detail 6 months after the application of a calcium/barium hydroxide nanoparticle mixture. Pictures a, b, d and e are readapted with permission from Giorgi et al. 2010b. (Copyright 2010 Wiley-VCH Verlag GmbH & Co. KGaA, Weinheim)

Recently, Poggi et al. (2014) reported an alternative process for the synthesis of $Ca(OH)_2$ nanoparticles, i.e. through an alcohol-thermal reaction starting from bulk metal and short chain alcohols such as ethanol and 1-propanol. The synthesis is carried out at high temperature and pressure, and consists of two steps: first calcium metal is oxidized by the alcohol so that an alkoxide is formed; then this intermediate is hydrolyzed leading to the formation of the hydroxide nanoparticles already dispersed in the appropriate solvent for practical application. The process does not require any purification step, and leads to the production of concentrated dispersions (35 g/L or higher), which favors the up-scaling of the production. The particles dispersed in ethanol have a bi-modal distribution with a population centered at 80 nm and a minor population centered at 220 nm. The particles in 1-propanol tend to form clusters of about 260 nm. In both cases the process leads to the production of highly crystalline particles, which can be used for consolidation. Similar nanodispersions are also useful for pH control, and have been registered in 2014 under the trademark Nanorestore Paper®, available on request from CSGI.

As mentioned in Sect. 2.1, other research groups have also produced and used formulations of calcium hydroxide nanoparticles for the preservation of immovable works of art.

Daniele et al. (2008, 2010) prepared calcium hydroxide nanoparticles either following the original synthetic process in homogeneous aqueous phase by Ambrosi et al. (2001a) or introducing modifications of the method. These applications regard stone conservation. The same author also investigated the interaction at room temperature between $Ca(OH)_2$ nanoparticles obtained with two different methods and silica fume (Daniele et al. 2013). Results indicated that by reducing particles dimensions the formation of calcium silicate hydrate phase (layers with wrinkles and rough-edged surfaces) occurred already after 7 days of hydration, while the complete consumption of free calcium hydroxide took place after 28 days. Moreover, Volpe et al. (2013) patented a process for the synthesis of $Ca(OH)_2$ nanoparticles by means of ionic exchange resins.

Ziegenbalg (2008) provided an overview of the commercial product CaLoSiL® (by IBZ-Salzchemie GmbH & Co. KG, http://www.ibz-freiberg.de/en), produced since October 2006, and of its use for the strengthening of carbonatic stones (see also Sect. 2.2.2). In this case, calcium hydroxide nanosols are synthesized directly from alcohol solution. The obtained particles have dimensions of 50–250 nm, and are dispersed in short chain alcohols such as ethanol, 1-propanol and 2-propanol, with a concentration range of 5–25 g/L. D'Armada and Hirst (2012) reported the use of this product for the consolidation of historically important medieval wall paintings at All Saints' Church (Little Kimble, Buckingamshire). In particular, the paintings were quite stable, but the walls exhibited many later plaster repairs that were in bad conditions and affected by efflorescence. Consolidation was to be carried out while allowing evaporation of moisture, therefore conservators selected a compatible inorganic material ($Ca(OH)_2$ nanoparticles) to strengthen the repairs. The particles were used to fill fine cracks that were adjacent to delaminating plaster and paint layers in need of stabilization. On average, ten applications were carried out in each area that required consolidation, obtaining positive effects without any visible change to the surface. The paper also includes several practical guidelines that are consistent with those found in earlier works on $Ca(OH)_2$ nanoparticles, mentioned in the previous paragraphs (application procedures will be further detailed in Sect. 2.4). According to the authors, nano-lime in ethanol seemed to provide better consolidation than in 2-propanol, probably due to better penetration of particles in ethanol. In fact, the authors stressed the importance of gathering more field-testing and of disseminating the obtained results. Application of CaLoSiL® products to wall paintings were also reported by Dahene and Herm (2013) in a work where the calcium hydroxide nanoparticles were used to flatten and consolidate delamination of the pictorial surface of ancient Roman *frescoes* at the "Herculaneum" excavation site (Ercolano, Napoli, Italy).

2.2.2 Formulations for Stone

Similar considerations as those made in the previous section apply also for the consolidation of stone. Traditional conservation materials include:

- Synthetic organic polymers, which can undergo degradation and lack physico–chemical compatibility with stone.
- Akoxysilanes or alkylalkoxysilanes (e.g. methyltrimethoxysilane, MT-MOS, and tetraethoxysilane, TEOS), which have poor affinity with calcite and exhibit cracking during shrinkage (Ferreira Pinto and Delgado Rodrigues 2008).

The cracking issue has been recently tackled, for instance by adding nano-sized silica particles and functional silanes that reduce the capillary force developed during the evaporation of solvents (Kim et al. 2009; Son et al. 2009), or by designing a consolidant synthesis where the sol–gel transition occurs in the presence of a surfactant, so to avoid cracking as the gel dries within the stone pores (Mosquera et al. 2008). Miliani et al. (2007) studied particle-modified silica consolidants (PMC) filled with titania, alumina and silica nanoparticles. Results indicated that PMC offer improved properties with respect to unfilled ethyl silicate consolidant, especially for what concerns protection from salt crystallization and reduced risk of cracking during thermal cycles.

Alternatively, research work has been carried out in the last 10–15 years on the formulation and use of nanoparticles of calcium hydroxide in alcohols, as discussed for wall paintings.

Ambrosi et al. (2001b) tested $Ca(OH)_2$ nanoparticles in 1-propanol both on laboratory samples simulating mortars with flaking surface and on real historical stone. In the first case, the treatment increased the surface compactness and lead to the formation of intergrain junctions as shown in Fig. 2.12. Application on calcareous stone ("Pietra di Nanto") constituting the external walls of the St. Margherita Abbey in Vigonza (Padua, Italy) lead to the consolidation of the surface layers, as indicated by scotch tape test. Treatment with nanoparticles was also carried out on the external walls of the apse of the St. Prisca in Aventino church (Rome), where brick surfaces had to be protected by a thin surface protection layer (50–100 μm).

Croveri et al. (2004) assessed the effectiveness of $Ca(OH)_2$ nanoparticles in 2-propanol for the consolidation of samples of *Globigerina* limestone, a biocalcarenite stone typically found in Malta. The authors compared the method with the application of two well-established materials for stone conservation, i.e. barium hydroxide and ammonium oxalate, both applied as aqueous solutions using cellulose poultices. The assessment included absorption and water vapor permeability tests, colorimetric analysis, SEM-EDX to investigate changes in the morphology and pore distribution of the treated samples, analysis of surface area and evaluation of consolidation through a micro-drilling system. Results indicated that treatment with nanoparticles produced the best effects in terms of surface consolidation. The physico–chemical properties of the substrate were respected, and alteration of the pore structure was small, which is essential to grant transpiration in seaside environments exposed to marine aerosol. Dei and Salvadori (2006) used the formulation in 2-propanol to counteract surface powdering and flaking on real samples: a soft calcareous sandstone (*Gallina* stone) coming from the façade of a historical building in Mantova (Italy), and a marly compact limestone (*Alberese*) coming from the

Fig. 2.12 Optical microscopy images under glazing light of aerial mortar surfaces: **a** decohered surface before application of $Ca(OH)_2$ nanoparticles; **b** the same surface consolidated by $Ca(OH)_2$/1-propanol nanoparticles dispersions. (Reprinted with kind permission from Springer Science+Business Media: Ambrosi et al. 2001b, Fig. 2.4.)

Santa Maria Church in Impruneta (Florence, Italy). The treatment was carried out through immersion for 8 h, drying for 1 h and quick washing with water to remove the consolidant excess from the surface, preventing the formation of white hazes. The consolidation effects were assessed after complete carbonation, in laboratory hygrometric conditions, 3 weeks later. The application resulted in consolidation both at a superficial level (re-aggregation of the powdering surface) and at a deeper level (decreased absorption of water by the substrate), without any detrimental effect on the samples physico–chemical properties. Baglioni et al. (2006, 2014) applied the same formulation to consolidate limestone in the archaeological site of Calakmul, Mexico. The paper provides details of the application procedure, which will be resumed in Sect. 2.4.

Lopez-Arce et al. (2010) studied the application of commercial calcium hydroxide nanoparticles dispersion in 2-propanol (Nanorestore®) to consolidate dolostone samples from the Madrid area, which are typically used in historical buildings. The stone samples were impregnated through a capillary tube and then introduced in climatic chambers to monitor the consolidation effects at different relative humidity (RH) values. The effectiveness of the treatment was assessed with several analytical techniques: Environmental Scanning electron Microscopy (ESEM-EDS, spectrophotometry, capillarity, water absorption under vacuum, ultrasound velocity,

Nuclear Magnetic Resonance (NMR, imaging and relaxometry) and Optical Surface Roughness. The product was also studied separately, by dropping it into plastic cups that were placed into climatic chambers, and the development of different calcium carbonate phases was monitored over 20 days. Results indicated that at RH = 75 % the consolidation process is favored. Calcium hydroxide (portlandite) undergoes fast transformation into vaterite ($CaCO_3$), monohydrocalcite ($CaCO_3 \cdot H_2O$) and calcite ($CaCO_3$), with the formation of larger crystals (from nano- to micron-size) and ultimately the treatment improves considerably the physical and hydric properties of the sample. Consolidation was achieved also at RH = 33 %. However in such conditions the application of the consolidating product, besides favoring calcite re-crystallization, leads also to the dissolution and fracture of the dolomite crystals ($CaMg(CO_3)_2$). Moreover in this case the conversion of calcium hydroxide to carbonate is slower and results in smaller particles that reach micron-size after 20 days due to an agglomeration process. Overall, the consolidating product does not alter the stone color. The substrate porosity decreases and the ultrasound velocity increases after treatment, indicating effective consolidation. Both these effects and the homogeneous penetration of the nanoparticles dispersion are enhanced for higher porosity samples. NMR analysis showed the filling of big pores and of cracks. Moreover, it was demonstrated that high porosity dolostone treated with nanoparticles and exposed at RH 75 %, room T and room CO_2 concentration for 20 days, showed a higher improvement of the physical and hydric properties even in the long term, i.e. 18 months after treatment, as compared to stone treated at RH 33 % (Lopez-Arce et al. 2013).

As mentioned before, the consolidation of dolomite stone can be achieved through the application of a mixed $Ca(OH)_2$ and $Mg(OH)_2$ nanoparticles dispersion in propanol (see Fig. 2.13), as in the case of the consolidation of "Pietra d'Angera" (Angera stone), a high-porosity dolomite stone widely used in Northern Italy for historical and monumental buildings. The mixed formulation grants the best compatibility with the stone, leading to durable consolidation (Chelazzi et al. 2013).

Even though in principle calcium hydroxide formulations have been designed for the consolidation of carbonate stone such as limestone, some authors have used $Ca(OH)_2$ formulations to consolidate sandstone, with good results (Rodriguez-Navarro et al. 2013). In particular, commercial nanoparticles (Nanorestore®) or carbide lime putty resulted in the formation of calcite or vaterite that bind quartz grains. Both 2-propanol and ethanol were used to disperse nanoparticles, the use of ethanol resulting in better performances for stone consolidation.

The commercial formulation CaLoSiL® has also been used for stone consolidation with positive results, and several authors focused on the best conditions and applicative protocols to maximize consolidation effects (Ziegenbalg 2008; Maryniak-Piaszczynski et al. 2008; D'Armada and Hirst 2012; Dahene and Herm 2013).

Ciliberto et al. (2008) proposed the synthesis of $Sr(OH)_2$ (strontium hydroxide) nanoparticles through homogeneous phase reaction (bottom–up approach), and their use as consolidant for stone, mortars and wall paintings. Nearly rounded crystalline particles of about 30 nm were obtained and dispersed in 1-propanol for application to porous bio-calcarenite (Lecce stone). The conversion of $Sr(OH)_2$ into

Fig. 2.13 (*Top row*) The historical "Ca' Granda" building in Milan (Italy). On the *right*, the restoration site is shown, where nanoparticles were tested on stone. (*Bottom row*) Application of nanoparticles on degraded stones surface in Ca' Granda, Milan, Italy. On the ***left***, the untreated and powdering portion of Pietra D'Angera. On the ***right***, application of consolidating nanodispersion by means of a brush. (Reprinted from Chelazzi et al. 2013. Copyright 2013, with permission from Elsevier.)

$SrCO_3$ provides the consolidation of the substrate. Similarly to barium hydroxide, strontium hydroxide nanoparticles react with gypsum to form an insoluble and non-detrimental sulfate ($SrSO_4$ in this case) and calcium hydroxide. This could be useful for the consolidation of sulfate-polluted substrates when the removal of salts (e.g. with the Ferroni method) can not be carried out (see Sect. 2.3).

In fact, Licchelli et al. (2014) assessed the use of $Ca(OH)_2$ nanoparticles obtained in ethanediol (see Sect. 2.2.1) and of $Sr(OH)_2$ nanoparticles for the consolidation of Lecce stone. Particles dispersions in 2-propanol were applied either by brushing or by capillary absorption. The authors concluded that application by brushing results in a deeper penetration and a more homogeneous distribution of the newly formed carbonate into stone. The "breath" of the original material was not dramatically altered, since water vapor permeability decrease was in any case lower than 30 %. In particular, stone specimens were treated with the nanoparticles dispersions and, once dry, were exposed to cycles of immersion in sodium sulfate solution, drying in oven at 60 °C and cooling at room temperature, to simulate damage from salt crystallization. The samples weight loss was recorded after the aging cycles. Results indicated that treatment with either $Ca(OH)_2$ or $Sr(OH)_2$ nanoparticles increased the resistance of stone to aging.

Finally, other researchers (Favaro et al. 2008; De Zorzi et al. 2009) proposed an alternative approach where calcium alkoxides (in methanol solution) impregnate porous substrates and then react with atmospheric moisture and CO_2 to form calcium carbonate *in situ*, which exhibits nano-structuration. The time needed for complete carbonation depends on thermohygrometric conditions and the thickness of the alkoxide particles, and ranges from 2–4 to 35–45 days. Eventually, a calcium carbonate film is produced, which adheres to the substrate grains without producing cracks, and binds the grains together. Tests on aged marble showed that the treatment results in stone consolidation, highlighting the good potentiality of this method. The main disadvantage is due to the low amount of carbonate deposited in each application, due the low solubility of alkoxides. Research is in progress to overcome this limitation. Recently these systems were used on marble specimens, and the consolidating effect was assessed through ultrasonic velocity measurements (Ossola et al. 2012). The consolidation appears to be related to the alkoxides solubility and stability in solution.

2.3 Application Field

Both laboratory research work and field applications in the last decade clearly pointed out that the effectiveness of wall paintings and stone consolidation treatments is affected by several factors, including:

- Environmental conditions (e.g. temperature, relative humidity)
- Characteristics of the consolidant (particles size, solvent, aging time)
- Substrate physico–chemical properties (chemical composition, porosity, etc.)
- Presence of pollutants (e.g. salts), degradation products and other compounds from previous restoration interventions, such as synthetic polymers
- Applicative procedure (this fundamental point will be addressed in the following section)

The literature reports several studies that contributed to outline the application field of nanoparticles dispersions for consolidation purposes, taking into account the aforementioned factors.

The rate of carbonation of slaked lime, including nano-sized lime, is lower at low temperatures, and at 4 °C or below the process is inhibited (D'Armada and Hirst 2012).

The effects of relative humidity (RH) have been deeply investigated, and research work is still in progress. The importance of RH in the carbonation of calcium hydroxide nanoparticles is not surprising, if we consider that the transformation involves the following processes:

- The diffusion of CO_2 in the gas/water interface.
- The dissolution of CO_2 in water and the formation of H_2CO_3 (carbonic acid).
- The dissociation of carbonic acid into CO_3^{2-} (carbonate ion) and HCO_3^- (bicarbonate ion).
- The dissociation of $Ca(OH)_2$ into Ca^{2+} (calcium ion) and OH^- (hydroxyl ion).
- Finally, the precipitation of $CaCO_3$ (calcium carbonate) formed by calcium ions and carbonate ions.

It is known that high RH values (>75 %) speed up the carbonation process, due to the adsorption of multilayers of water onto the solid $Ca(OH)_2$ (Beruto and Botter

2000). However, as RH approaches saturation (e.g. RH ≥98 %) the carbonation process is inhibited, due to the Le Chatelier principle (the carbonation reaction produces water molecules, besides calcium carbonate, starting from calcium hydroxide and carbon dioxide). Therefore, the presence of a macroscopic film of water over the $Ca(OH)_2$ layer will delay carbonation. In fact, lime putty is stored and aged under bulk water.

Besides the process rate, another important aspect regards the formation of different calcium carbonate polymorphs, i.e. different forms or crystal structures of calcium carbonate. The main calcium carbonate polymorphs are:

- Three anhydrous crystalline forms: calcite, aragonite and vaterite
- Two crystalline hydrated forms: monohydrate (monohydrocalcite, MHC) and hexahydrate (ikaite)
- Amorphous calcium carbonate hydrate (ACC)

It is important to notice that the formation of different carbonate forms results in different physical properties of the treated artifact. In fact, the various carbonate forms have different stability. For instance, vaterite is more soluble than calcite; moreover, its crystalline structure (polycrystalline spherulites) is different than that of calcite, which might hinder the formation of a coherent bond between vaterite (coming from the treatment) and calcite-based substrates such as lime mortars, plasters and limestone. These aspects are important in the short-term, but in time vaterite will dissolve and re-precipitate as calcite, which is stable, granting long-term positive effects. It has been reported that the vaterite-to-calcite conversion process can be long (e.g. several months) even at high RH (80 %) (Rodriguez-Navarro et al. 2013). Despite these limitations, the same authors found that vaterite shows indeed some effectiveness in consolidating stone.

Lopez-Arce et al. (2011) have studied the commercial formulation Nanorestore® and have clearly demonstrated that both the carbonation rate and the formation of different carbonate polymorphs depend strongly on RH. The transformation of calcium hydroxide into carbonate was monitored over 28 days. At RH values of 75–90 % the carbonation process is faster and calcium hydroxide is completely transformed into carbonate within 7 days, as water catalyzes the initial formation of ACC and MHC. ACC is unstable and in the presence of water it quickly converts into vaterite and/or calcite. The MHC content decreases with time too, as it converts into anhydrous polymorphs. The particles dispersing medium plays also a fundamental role, connected with RH: higher RH delays the evaporation of alcohol molecules that stabilize vaterite and aragonite, and indeed the relative amount of calcite formed after 28 days is either 11 % (RH = 75 %) or 37 % (RH = 90 %). Overall, at higher RH values the formed carbonate particles are larger and exhibit higher crystallinity.

On the other hand, at RH values of 33–54 %, the carbonation process is slower and the hydroxide–carbonate conversion yield is much lower. The calcium hydroxide leftover after 28 days is 84 % (RH = 33 %) or 36 % (RH = 54 %). At RH = 33 % alcohol will evaporate faster and eventually (28 days) the small amount of calcium carbonate is equally in the form of calcite and MHC. At 54 %, the formed carbonate is mainly vaterite, and no calcite is detected. Overall, at lower RH values the carbonate particles are smaller and exhibit lower crystallinity.

Gomez-Villalba et al. (2011, 2012) found similar results for the CaLoSiL® commercial product (in 2-propanol). Other authors compared the product in 2-propanol and ethanol, the latter showing higher effectiveness for consolidation purposes (D'Armada and Hirst 2012).

Rodriguez-Navarro et al. (2013) further investigated the role of alcohol in the carbonation kinetics. According to the authors, the contact of particles with alcohol (e.g. during the storage of dispersions) results in the replacement of $Ca(OH)_2$ by calcium alkoxides. For instance in the case of nanoparticles dispersed in ethanol, the alkoxide formation is described by the following chemical equation:

$$Ca(OH)_2 + 2CH_3CH_2OH \rightleftarrows Ca(OCH_2CH_3)_2 + 2H_2O$$

The replacement reaction seems to preserve the hexagonal platelet-like shape of the particles, and the reaction yield depends on the reactivity of the particles, i.e. smaller particles with high surface area and particles with lattice defects will be more reactive. After their deposition on the artistic substrate, the exposure of nanoparticles to high RH values (80 %) results in fast hydrolysis of alkoxides, and both the carbonation rate and yield are increased. The stabilization of vaterite is favored, probably due to the alcohol molecules released during the alkoxides hydrolysis (i.e. the reverse of the aforementioned chemical reaction).

Conversely, when the contact time between nanoparticles and alcohol is reduced (e.g. short storage before use), the alkoxide amount is lower. In this case, after deposition of the particles carbonation takes place at a slower rate, the process yield is decreased, but the formation of calcite will be favored.

Eventually, the authors suggest that long contact (storage) time, for instance much bigger than 2 months, can be preferred whenever the dispersion must be used for pre-consolidation treatments, where fast and high-yield carbonation is desired. Otherwise, fresh dispersions could be selected for long-term effects.

For what concerns the size of the nanoparticles, it must be noticed that size reduction is not the only valid criterion for consolidation treatments. In fact, the best effects are expected when the particles size distribution matches that of the substrate's pores. Therefore, when both small and large pores are present, one can either use a formulation where particles exhibit a bimodal size distribution, or alternatively different formulations can be used (obviously the two formulations must be compatible with each other and with the substrate).

As we discussed in the previous sections, the chemical composition of the applied materials (nanoparticles) should mirror as much as possible that of the substrate. The use of mixed $Ca(OH)_2$ and $Mg(OH)_2$ nanoparticles for the consolidation of dolomite stone is an example of adapting the consolidant composition to that of the substrate in order to maximize the results. According to this principle, calcium hydroxide nanoparticles can be considered as the ideal consolidation material for mainly carbonate-based stones. However, in some cases good consolidation effects have been observed also on sandstone (see above). Rodriguez-Navarro et al. (2013) provided insightful comments to explain this behavior: "First, calcium carbonate is a common cement in many sandstones and strongly contributes to their durability. Second, it has been shown that building sandstones, as well as loose quartz sandy soils, can be efficiently consolidated by $CaCO_3$ (vaterite and/or calcite) produced by bacterial biomineralization. Ultimately, the applied treatment mimics the natural or bacteriogenic carbonate cementation of sandstones or quartz sand, and effectively improves the physical–mechanical resistance of such a siliceous material". The authors also noted that, according to the literature, amorphous or poorly crystalline gel-like calcium silicate might form owing to the interaction of calcium hydroxide and SiO_2 grains, in the presence of water (e.g. moisture in the pores). The formation of this phase in conditions of high alkalinity would grant a "chemical continuity" between carbonated lime and silicates (Armelao et al. 2000). In fact, tests recently showed that consolidation of degraded stone can be enhanced by carrying out a pre-treatment with $Ca(OH)_2$ nanoparticles followed by treatment with ethyl silicates. However, further study is necessary since for pure sandstones the use of consolidants based on the carbonation of $Ca(OH)_2$ might eventually lead to the introduction of calcium salts such as gypsum (D'Armada and Hirst 2012).

It is known that the presence of salts in the porous matrix of wall paintings or stone can affect the consolidation process (Giorgi et al. 2010b). Highly soluble salts, such as sodium sulfate, dissolve in water within the pores, and the free sulfate ions precipitate with calcium ions coming from the dissolution of the nanoparticles (the solubility of calcium hydroxide is low but can not be completely neglected), forming gypsum. Both the partial consumption of calcium hydroxide and the formation of gypsum hinder the consolidation process, with detrimental effects. In this case, a mixed $Ca(OH)_2$ and $Ba(OH)_2$ nanoparticles dispersion can be applied: barium hydroxide reacts with gypsum forming the insoluble barium sulfate (which is not detrimental), calcium hydroxide and water. The mixed nanoparticle dispersion is also useful when it is not possible to carry out the Ferroni method previously to the consolidation treatment, for instance when the pictorial surface of a wall painting is extremely fragile and degraded.

2.4 Application Procedures

The aim of this section is to provide practical guidelines for the application of dispersions of $Ca(OH)_2$ nanoparticles in alcohols such as 1-propanol, 2-propanol and ethanol. In particular, these guidelines are to be used for the application of $Ca(OH)_2$ nanoparticles dispersions as designed by CSGI (Consorzio Interuniversitario per lo Sviluppo dei Sistemi a Grande Interfase, Center for Colloid and Surface Science), University of Florence, and commercialized as Nanorestore®. Different formulations might behave differently, so adaptations of the indications here provided should be considered case by case.

Dispersions of $Ca(OH)_2$ nanoparticles were initially designed to fix pulverization, and to re-adhere flaking and detached parts on the surface of *fresco* paintings.

However, this methodology has proved effective also on carbonate-based *secco* paintings, and to some extent on stone, as described in the previous sections. The proposed method is based on the theoretical and practical knowledge gathered in the last decades through both research work and field activity in collaboration with conservators. It is fundamental to recall that each case study exhibits specific characteristics and issues; therefore the practical guidelines introduced here should be considered and applied critically, and not as recipes. The best approach is always the collaboration and interactive feedback between conservation scientists and restorers, to grant the flexibility and adaptability of the methods, minimizing (or avoiding) drawbacks and maximizing the results.

Before consolidation with nanoparticles, it is fundamental to carry out diagnostic analyses to precisely characterize the artifact's composition and degradation status. Environmental conditions, such as temperature and relative humidity oscillations, should be known. Moreover, typical issues such as building's structural problems and sources of moisture must be dealt with previously to any consolidation treatment, and general preventive conservation measures must be taken, in order to set the best conditions for practical interventions. Records of the artifact's history are mostly important as a support to diagnostics, so to collect complete information on previous interventions (application of synthetic polymers, protectives, etc.) and conservation issues (presence of salts, pollutants, etc.). Based on this information, scientists and conservators can change some of the features of the intervention, such as the order of application of materials, the quantity of materials to be applied, the application method and time, etc. The main criterion is the respect of the artifact's physico–chemical properties. Preliminary consolidation tests must be carried out to understand as much as possible the effectiveness of the intervention, and possible drawbacks.

After this fundamental premise, the general procedure for the consolidation treatment can be described as follows:

- If the artifact does not exhibit sulfate pollution, the consolidation can be carried out directly with $Ca(OH)_2$ nanoparticles dispersions.
- *In case of sulfate pollution*, the removal of sulfates must first be carried out using the *first step* of the Ferroni method (application of ammonium carbonate solution, see next Section). The artifact's consolidation is then carried out through the application of $Ca(OH)_2$ nanoparticles dispersions.
- *In case of large amounts of sulfates*, the first step of the Ferroni method is carried out to remove the sulfates as much as possible, and then the artifact consolidation can be carried out through the application of mixed $Ca(OH)_2$ and $Ba(OH)_2$ nanoparticles dispersions (see Sect. 2.4.2 for further details).

2.4.1 The Ferroni Method

The presence of sulfates is one of the main degradation sources for immovable works of art such as wall paintings and stone, as described in Sect. 2.1. Moreover, high-solubility sulfates such as sodium sulfate (Na_2SO_4) hinder the consolidation effectiveness of $Ca(OH)_2$ nanoparticles, as explained in Sect. 2.3.

At the end of the 1960s, Enzo Ferroni developed a method for the removal of sulfates and the consolidation of flaking surfaces. The method has been successfully applied in numerous occasions, and is well accepted and used worldwide.

As a preliminary procedure, and only if the conservation status of the artifact allows it, the surface of the work of art can be cleaned gently (e.g. with distilled water) to remove dust and impurities. Great care must be taken to avoid the removal of original artistic materials.

The Ferroni method (also known as the barium method or the Florentine method) is a 2-step method. The first step consists in the application of ammonium carbonate; the second step is the application of barium hydroxide.

In the first step, ammonium carbonate, $(NH_4)_2CO_3$, is dissolved in distilled water up to a final concentration of 3–10%. Higher concentrations are useful for the removal of bigger amounts of sulfates. The ammonium carbonate solution is then loaded into a cellulose poultice, which is applied (over a veil of Japanese paper) on the artifact surface. Arbocel® BC200 or Arbocel® B40 are typically used. The poultice must be impregnated with the aqueous solution, but not dripping, so to stick to the surface. The Japanese paper grammage is relevant: typically a 9 g/m^2 is used, but it also depends on the application. The poultice thickness can vary from 5 mm to 2 cm. The poultice must be compact and well adhered to the artifact surface, without empty gaps in between. The adhesion of the poultice to the surface is normally granted by applying a small pressure, upon which no dripping must take place.

Ammonium carbonate transforms the fairly soluble gypsum (solubility: 2.4 g/L at 20 °C) into the highly soluble ammonium sulfate (744 g/L at 20 °C), according to the following ion-exchange reaction:

$$CaSO_4 \cdot 2H_2O + (NH_4)_2\,CO_3 \rightleftarrows (NH_4)_2\,SO_4 + CaCO_3 + 2H_2O$$

Being highly soluble, ammonium sulfate is mostly absorbed by the poultice. In the reaction, calcium carbonate precipitates as a powdery filler with scarce mechanical properties.

There is no standard application time, and if the status of the artifact allows it the poultice can be applied until it dries completely, also depending on the environmental conditions (T, RH).

After the application time, the poultice is removed but the Japanese paper is left on the surface for the washing step. In fact, it is necessary to wash the surface in order to remove any salt residues. This is usually accomplished by applying a new cellulose poultice (still over Japanese paper) loaded with distilled or deionized water. The poultice is left onto the surface until it dries. It is advisable to repeat the washing process at least 2–3 times, using each time new poultices loaded with fresh water, so to remove all the salts. The presence of sulfates in the washing water can be checked each time, e.g. by measuring water's conductivity. However, the application of barium hydroxide in the second part of the method (as explained in the following paragraphs) will turn residual soluble sulfates into insoluble barium sulfate, which is not detrimental.

Finally, the Japanese paper is removed very gently, taking great care not to cause the detachment of fragile surface layers, such as flaking pigments. However, if the surface is extremely fragile and damaged, it is advisable to leave the Japanese paper and consolidate the surface using a barium hydroxide solution or calcium hydroxide nanoparticles. We recall that the consolidation effectiveness of $Ca(OH)_2$ nanoparticles is hampered by the presence of sulfates, which should be removed.

The main practical steps for the removal of sulfates with the Ferroni method are depicted in Fig. 2.14.

Before describing the second part of the method, it is essential to provide some further considerations on the application of ammonium carbonate.

Ammonium carbonate is typically applied when the presence of sulfates is visually evident in the form of veils or efflorescence, or when the salts are detected through diagnostic investigation, for instance through infrared spectroscopy. If scientific facilities for diagnostics are not accessible, the presence of the salts can be identified with the following procedure:

1. A small quantity of the efflorescence is collected from the surface of the work of art.
2. The collected sample is placed over a clean surface (resistant to acids) and a drop of hydrochloric acid, HCl, or nitric acid, HNO_3, both diluted in water (5–10 % v/v), is added to the sample. A drop of barium chloride aqueous solution ($BaCl_2$, 5 % w/w) is then added. If sulfates are present in the sample, a white precipitate is formed. The test result might be difficult to observe. In this case, the test can be carried out under a microscope: if sulfates are present in the sample, after the evaporation of water one should observe the formation of acicular crystals.
   Alternatively, the sample is put in a cleaned glass vial, 2–3 mL of distilled water is added and then hydrochloric acid or nitric acid are used to acidify the solution. Then a small quantity (a pinch) of barium chloride salt is added and the solution is stirred for a couple of minutes. The formation of a white precipitate indicates the presence of sulfates in the sample.

If the test is negative, the presence of sulfates should be double-checked using advanced diagnostic methods. If the test is positive, preliminary tests with ammonium carbonate must be carried out on marginal spots (5 × 5 cm), avoiding if possible drawings or figures, to determine the needed $(NH_4)_2CO_3$ concentration.

Whenever ammonium carbonate is applied on wall paintings, it is ***fundamental*** to identify the chemical composition of the pigments (and of their degradation products) *previously* to the application of the chemicals. In fact, the ammonium carbonate solution is alkaline and develops ammonia; therefore the solution can react

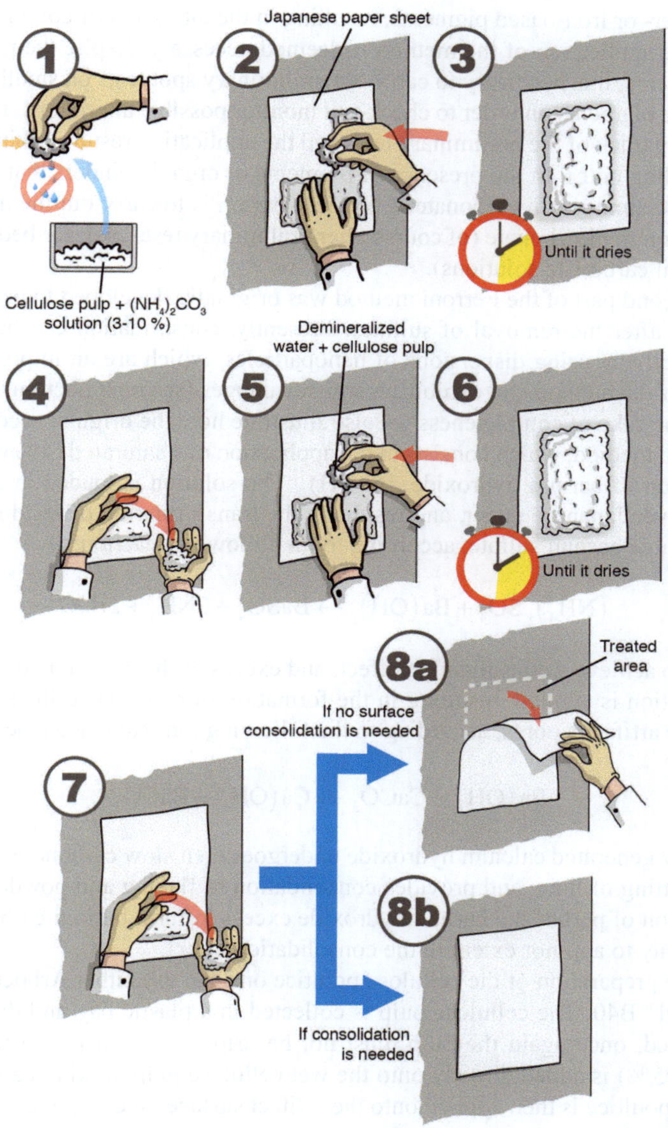

Fig. 2.14 The main steps of the first part of the Ferroni method: *1* the cellulose poultice is loaded with the ammonium carbonate solution; *2–3* the humid poultice is applied onto the wall painting, over a Japanese paper sheet, and left until it dries (if possible); *4* the poultice is delicately removed; *5–6* another poultice, soaked with demineralized water, is applied onto the Japanese paper sheet, until it dries; *7* Removal of the poultice; *8a* If the surface is not in need of consolidation, then the Japanese paper sheet can be removed; *8b* However, if the surface needs consolidation, it is advisable to leave the Japanese paper on and consolidate the surface using a barium hydroxide solution (i.e. second part of the Ferroni Dini method) or calcium hydroxide nanoparticles. (Image by Michele Baglioni)

with copper- or iron-based pigments, resulting in the alteration of colors and tones. In case the application of the method is deemed necessary despite the presence of such pigments, it is necessary to carry out preliminary spot tests on small areas that contain the pigments, in order to check and monitor possible alterations. It is important to keep track of the preliminary tests and the application results, taking pictures and recording notes. In the presence of pigments or organic binders that are highly sensitive to ammonium carbonate, a possible option is to carry out the method using ammonium bicarbonate (of course after preliminary tests as described above for ammonium carbonate solutions).

The second part of the Ferroni method was originally developed to provide consolidation after the removal of sulfates. Presently, consolidation can be achieved more effectively using dispersions of nanoparticles, which are an improvement to this part of the method and exhibit several advantages (see next Section). Nonetheless, for the sake of completeness we also illustrate here the original second step of the Ferroni method, which consists in the application of a saturated (5 % w/w) aqueous solution of barium hydroxide, $Ba(OH)_2$. The solution is loaded in a cellulose poultice over Japanese paper, and results in the transformation of residual sulfates into insoluble barium sulfate, according to the following reaction:

$$(NH_4)_2 SO_4 + Ba(OH)_2 \rightarrow BaSO_4 + 2NH_3 + 2H_2O$$

In order to achieve a consolidation effect, and excess of the barium hydroxide saturated solution is applied, resulting in the formation of new slaked lime ($Ca(OH)_2$) within the artifact's pores, according to the following ion-exchange reaction:

$$Ba(OH)_2 + CaCO_3 \rightarrow Ca(OH)_2 + BaCO_3$$

The newly generated calcium hydroxide undergoes then slow carbonation, mimicking the setting of lime, and provides consolidation to flaking and powdering parts. Carbonation of part of the barium hydroxide excess to form barium carbonate contributes only to a minor extent to the consolidation effect.

For the preparation of the cellulose poultice one can use either Arbocel BC®200 or Arbocel® B40. The cellulose pulp is collected in a plastic bag and distilled water is added, once again the pulp must not be dripping. Then powdery $Ba(OH)_2$ (purity $\geq$ 95 %) is added directly onto the wet cellulose pulp, mixing carefully. The cellulose poultice is then applied onto the artifact surface, over Japanese paper. The paper grammage is indicatively 9 g/L, but different values can be used depending on the artifact's conditions. However, a too high grammage might prevent the adhesion of the poultice to the surface. The cellulose poultice is left on the surface until it is dry, then it is carefully removed. Japanese paper is then removed carefully.

Due to the alkalinity of the $Ba(OH)_2$ solution, similar considerations as those previously described apply for pigments sensitive to alkalis, such as copper- or iron-based pigments. Preliminary tests must be carried out and recorded as explained in the first part of the method.

The time gap between the application of the first part (ammonium carbonate) and the second part (barium hydroxide) is ca. from 1 week to 1 month, depending on the environmental conditions. Before applying barium hydroxide, tests on small marginal spots (5 × 5 cm) should be carried out to evaluate the powdering of surface layers and any possible drawbacks such as color or tone alteration etc.

The main practical steps of the second part of the Ferroni method are depicted in Fig. 2.15.

2.4.2 Application of Nanoparticles Dispersions

As illustrated in the previous sections, $Ca(OH)_2$ nanoparticles dispersions show a high consolidating power on powdering and flaking substrates based on calcium carbonate. Moreover, when plasters and painted layers exhibit losses, the particles will "intelligently" place in such gaps and consolidate them. Indeed, the formation of carbonate crystals inside gaps, cracks and fissures on the artifacts surface is a positive result of the consolidation treatment (see Fig. 2.16), since the formed calcium carbonate crystals act as "micro-groutings", bridging degraded parts together and improving the mechanical properties of the cracked/fissured surface. For wall paintings, the consolidated gaps can then be re-touched safely, or alternatively pigments could be added to the particles to provide a one-step consolidation and retouch.

As mentioned above, the application of $Ca(OH)_2$ nanoparticles is a significant improvement to the second part of the Ferroni method (which uses barium hydroxide aqueous solutions). This is due to several factors:

1. The consolidating power of $Ca(OH)_2$ nanoparticles is higher than that of barium hydroxide aqueous solutions, also due to the high reactivity of the nanoparticles.
2. The application of dispersions of particles in alcohols allows a better control of the consolidation treatment. For instance, the uncontrolled leaching or extraction of water-soluble compounds from the artifact is avoided. Moreover, an interesting aspect of using particles dispersions rather than alkaline solutions is that when alkaline compounds such as hydroxides are applied as poorly soluble solid-state particles, possible alterations of alkali-sensitive pigments are strongly reduced or completely avoided. Field activity has shown that calcium hydroxide nanoparticles can be used for the consolidation of iron- or copper-based pigments such as Azurite, with no detrimental effects. However, as a general criterion, preliminary applications on small spots (5 × 5 cm, possibly on marginal areas) are advisable, to monitor possible drawbacks and the effectiveness of the application.

We recall that if the artifact does not show sulfate pollution, the $Ca(OH)_2$ nanoparticles dispersions can be applied directly.

In the presence of soluble sulfate salts (Na_2SO_4, $MgSO_4$), the removal of sulfates is carried out using the *first part* of the Ferroni method (application of ammonium

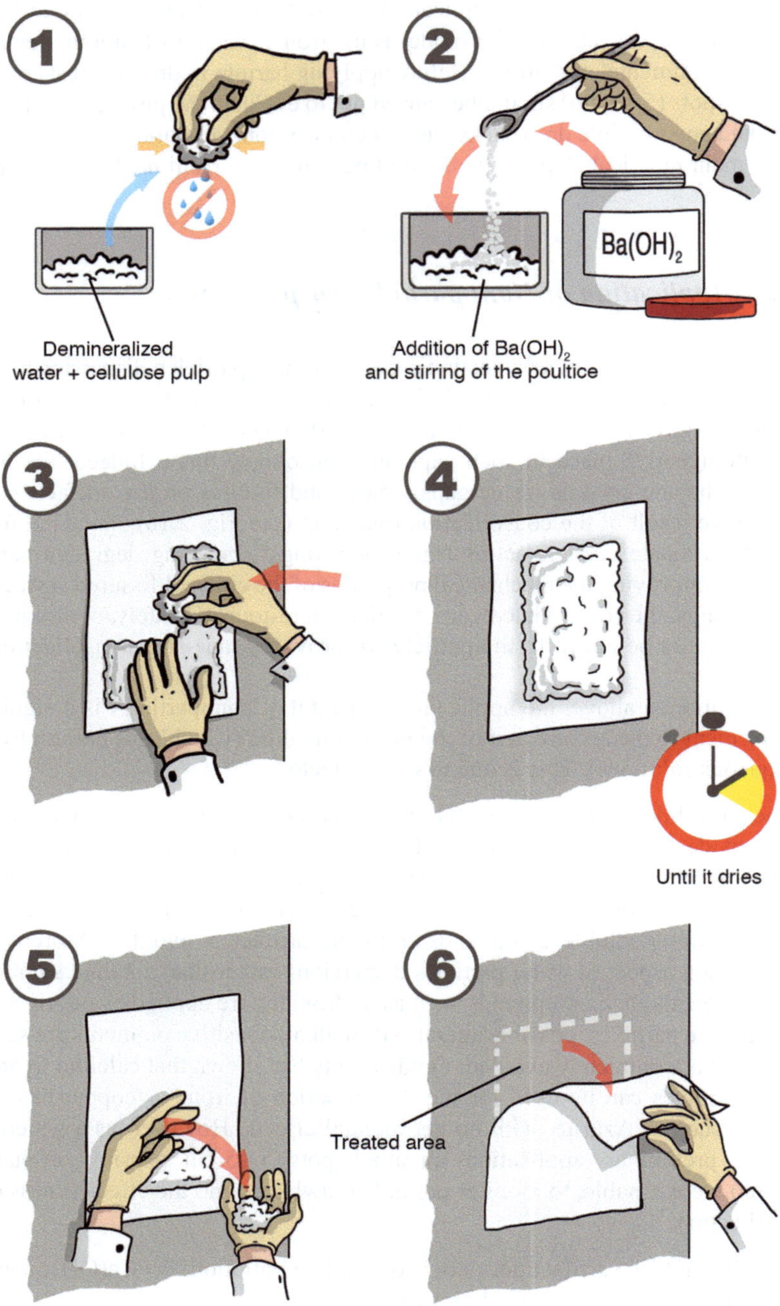

Fig. 2.15 The main steps of the second part of the Ferroni method: *1* demineralized water is added to cellulose pulp in a bag or container, until the pulp is humid (not dripping); *2* $Ba(OH)_2$ is added to the humid pulp (the amount of $Ba(OH)_2$ needed is 5% w/w of the water added at point *1*; *3–4* the humid poultice is applied onto the wall painting, over a Japanese paper sheet, and left until it dries (if possible); *5* the poultice is delicately removed; *6* removal of the poultice and sheet. (Image by Michele Baglioni)

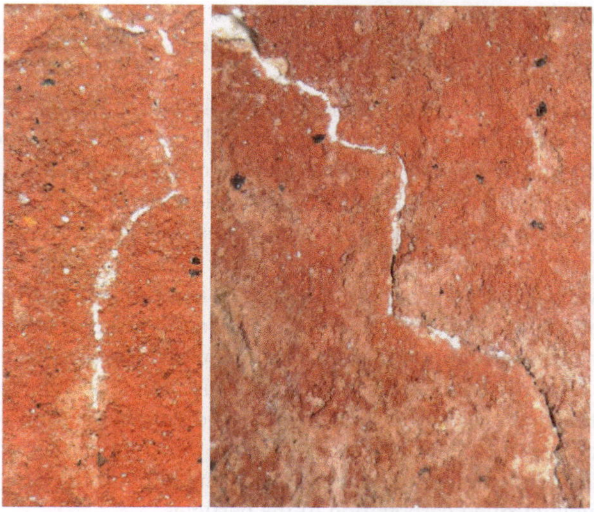

Fig. 2.16 Calcium carbonate crystals formed inside gaps, cracks and fissures on the artifacts surface, following treatment with $Ca(OH)_2$ nanoparticles. The formed calcium carbonate crystals act as "micro-groutings", bridging degraded parts together and improving the mechanical properties of the cracked/fissured surface. Eventually the consolidated white micro-groutings can be retouched with the appropriate pigment, or the pigment can be directly added to the nanoparticle dispersion.

carbonate solution). Then, the artifact's consolidation is carried out through the application of $Ca(OH)_2$ nanoparticles dispersions.

In case of large amounts of sulfates, the first part of the Ferroni method is carried out to remove the sulfates as much as possible, and then the artifact consolidation can be carried out through the application of mixed $Ca(OH)_2$ and $Ba(OH)_2$ nanoparticles dispersions. (If the artifact surface is strongly degraded and fragile, a pre-consolidation treatment with nanoparticles is useful to provide strengthening so to allow a safer removal of sulfates. Mixed calcium and barium hydroxide nanoparticles have proven to be useful also in this case).

As a general applicative criterion, it is fundamental to keep track of the environmental conditions (T, RH) before, during and after the application of the particles. Before the application, small tests with cotton swabs (wetted with distilled water) are useful to evaluate the degradation status of the artifact. All tests and applications must be documented with precise records so to allow the best control and development of the method.

The standard concentration of the Nanorestore® product is 5 g/L. However, lower concentrations, for instance 1 g/L, can be useful for application on low porosity substrates (also depending on the degradation of the substrate). As a general rule, it is better to apply the product several times (e.g. up to 10) at lower concentrations, rather than perform 1–2 applications at high concentrations. By working at lower concentration, in fact, the clustering of particles is reduced, penetration is enhanced, and the formation of white hazes on the treated surface is minimized or completely avoided. Depending on the substrate porosity and degradation status, it is also possible to apply dispersions of different concentrations, starting from the lower concentration up to the higher.

It is important to store the dispersions in firmly closed bottles, so to avoid any contact with air or moisture in order to prevent clustering of the nanoparticles (favored by water molecules) and transformation of hydroxide into carbonate (by CO_2). The latter effect would hinder the consolidating action of the particles. If stored in the above-mentioned conditions, Nanorestore® is a fairly stable dispersion but, especially at higher concentrations (5 g/L or more), deposits of nanoparticles can be observed in time. Thus, before use, the dispersion should be shaken energetically for a couple of minutes, and deposits need to be separated and discarded. If available, treatment with an ultrasonic bath can also be considered to favor the re-dispersion of particles' clusters. The product is typically considered "expired" after 12 months, however if well preserved (avoiding contact with water and CO_2 as explained above) the dispersions are stable for much longer periods. Dilution of concentrated dispersions must be carried out using anhydrous solvents to prevent clustering or uncontrolled carbonation of the particles.

The presence of coatings such as polymers, fixatives and glues, on the surface of the artifact might prevent particles penetration. The removal of the coating is a preliminary step before consolidation, however as explained before very fragile surfaces can pre-consolidated before and/or during the gradual removal of the coatings.

The nanoparticles dispersion should be applied on a dry surface (up to 50 cm × 50 cm). If the porous matrix to be consolidated (wall, stone, mortar) contains moisture and/or is wet, then a preliminary treatment with alcohol (the same as the dispersion's) is useful to dry the pores, granting a better penetration of the particles. For dispersions in 2-propanol, this preliminary procedure can be done in two steps using ethanol and 2-propanol. The extent of particles penetration depends largely on the substrate properties (pore size distribution, chemical composition, presence of moisture within the pores) and on the environmental conditions (T, RH, ventilation). Both researchers and end-users have reported penetration lengths varying from hundreds of microns, to a few millimeters, up to a few centimeters for highly porous and/or degraded substrates.

The application of the nanoparticles dispersion is normally carried out on the dry surface by brushing (using a flat brush) over a Japanese paper sheet. The paper must not exhibit any surface treatment (glues, sizing, etc.). A paper grammage of 9 g/m^2 is advisable, but it depends on the application. It is important that the Japanese paper sheet adhere to the surface roughness as it is pressed by the brushing action. High grammages might prevent the adhesion of the paper to the artifact surface. The brushing motion helps to re-adhere flaking parts and favors particles penetration. The dispersion can be applied several times, waiting a few seconds between each application, until it is "rejected" by the porous matrix, i.e. until it is evident that the matrix can not absorb any more dispersion, and any extra addition results in accumulation of the dispersion on the artifact surface. In case of less capillary active substrates particular attention should be paid on the slow and gradual application of the consolidant. The spraying method (see below) enables good control of the amount of material introduced into the substrate. Application by brush or syringe can produce bigger optical changes, and these methods are considered more suitable for more porous substrates. In any case, gradual application is highly recommended.

The product consumption is hardly predictable, as it largely depends both on the concentration of the applied formulation, and on the substrate properties (porosity, moisture content, etc.). On average, for wall paintings consolidation 1 L of product at 5 g/L is necessary to treat roughly 5–10 m^2 of pictorial surface with "average" porosity and decohesion. However, large deviations from this behavior have been reported.

After the application, it is necessary to put a cellulose poultice, soaked with distilled water, on the treated surface (still over Japanese paper). The poultice must be humid but not dripping. This operation prevents the formation of white hazes and favors the proper carbonation of the particles. The application of the poultice is particularly indicated when the environment Relative Humidity (RH) is below 70 %. The poultice is left until it dries, then it can be removed gently (along with the Japanese paper). If the Japanese paper exhibits adhesion to the surface, wetting with ethanol or propanol can be used to ease the gentle removal of the paper sheet.

In some cases white hazes can form, roughly within 20 min from the last application, due to several reasons, including high concentration of the particles, low porosity of the substrate, or a too dry environment that causes fast evaporation of alcohol, preventing penetration. Hazes can be removed by applying alcohol (the same as the dispersion) by brushing. If the haze persists, a cellulose pulp poultice soaked with deionized water must be applied for few hours. Moreover, light hazes usually disappear in 12–24 h when the treated materials are exposed to high relative humidity conditions (RH > 75 %).

As discussed in Sect. 2.3, carbonation of the nanoparticles depends largely on the environmental conditions, and also on the type of formulation used and on the substrate properties. Full carbonation might require, indicatively, from 1 to 4 weeks. It is advisable to assess the consolidation effectiveness approximately up to 1 month after the application.

Alternatively to brushing, nanoparticles can be applied using a syringe, or sprayed on the surface using a nebulizer and a mask to shield areas that are not to be treated. The syringe method is useful to inject the dispersion behind flaking scales and detachments that need to be re-adhered. Injection must be carried out carefully to avoid dripping of the dispersion on the artifact surface. Depending on the surface mechanical resistance, the operator can apply light pressure to facilitate the re-adhesion of the scale, as he/she injects the dispersion. Multiple injections might be necessary to achieve consolidation, which will be complete after full carbonation of the particles as explained above.

The spraying method is particularly indicated when dealing with a very fragile surface that could not withstand mechanical stresses, requiring thus a pre-consolidation treatment. The dispersions are sprayed on the surface of the poorly consolidated paint layer using an aerosol sprayer. End-users have reported optimal results using a Preval Sprayer (prod. Precision Valve Corporation, www.prevalspraygun.com), which has shown higher precision and evenness in dosage than hand-pumped sprayers, and was therefore considered a useful applicative tool. The sprayer is usually held ca. 10 cm from the surface, and the dispersion is applied until surface saturation. An application of excess dispersion is likely to produce a white haze, which can be removed as soon as it appears, applying a light spray of water. The dispersion can be applied again upon drying of the surface, e.g. within an hour. The appearance of white hazes is discouraged (and typically avoided) by using dispersions at lower concentration (from 1 to 3 g/L). The application of a wet cellulose poultice is then advisable as indicated before.

The main practical steps for the application of nanoparticles either using brush, syringe or sprayer have been depicted in Figs. 2.17, 2.18 and 2.19.

Finally, it is important to consider that in terms of ecotoxicological impact the $Ca(OH)_2$ nanoparticle formulations described in this Section can be considered very similar to the traditional solvents (e.g. 1-propanol, 2-propanol, ethanol), determining the same human health and environmental hazards of the organic solvents employed. Therefore the safe handling of the formulations involves the presence of good air ventilation, the use of suitable eye protection (lab glasses tested and approved under appropriate government standards, such as EN 166), no heating, and without sparks, open flames and hot surfaces. Handling also involves the use of hand protection (nitrile rubber gloves, minimum layer thickness: 0.2 mm), even if Skin Irritation Tests confirmed that the formulations are not irritant for skin.

For what concerns the application of barium hydroxide nanoparticles, the possible release from treated wall painting surfaces has been evaluated in order to confirm the safety of the method. Sulfate-polluted surfaces have been treated with a mixed $Ca(OH)_2/Ba(OH)_2$ nanoparticles dispersion, as in a real consolidation intervention. Then, the release of barium hydroxide from the treated substrates has been estimated simulating different leaching conditions (fog/light rain, heavy rain, flood). The maximum Barium concentration observed during the leaching tests (33 μg/L for the worst case scenario) is well below the maximum concentration limit (MCL) of 2 mg/L^{-1} of Barium in drinking water established by EPA (Environmental Protection Agency).

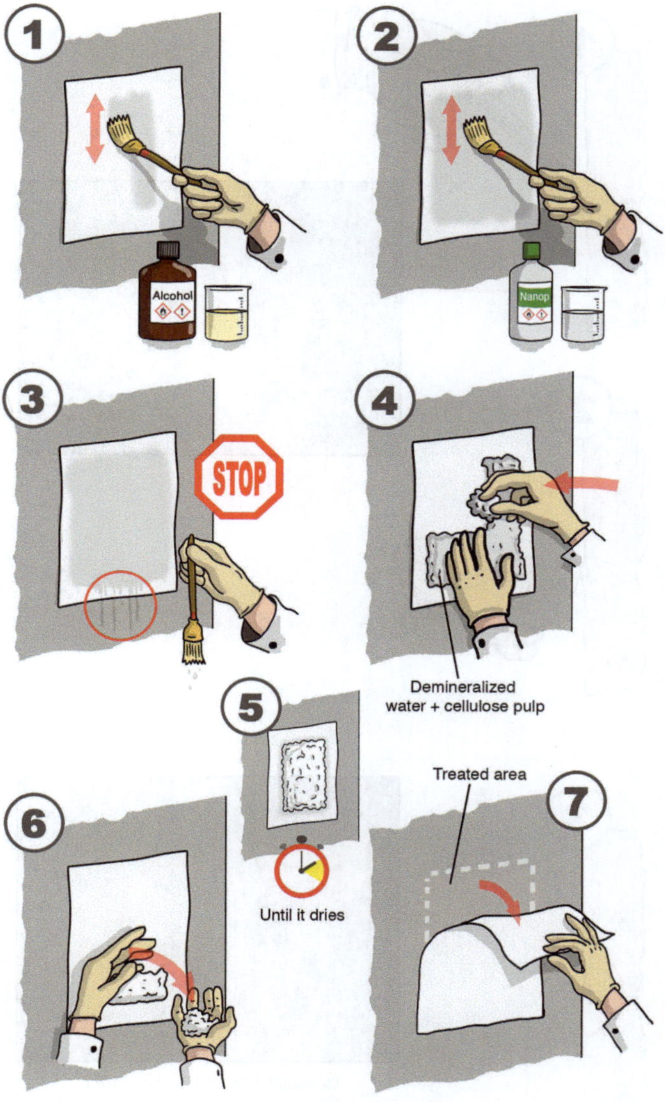

Fig. 2.17 Application by brushing of a dispersion of $Ca(OH)_2$ nanoparticles for the consolidation of a substrate. *1* If the porous matrix to be consolidated (wall, stone, mortar) contains moisture and/or is wet, a preliminary treatment with alcohol (the same as the dispersion's) is useful to dry the pores, granting a better penetration of the particles. *2* The application of the nanoparticles dispersion is carried out on the dry surface using a flat brush over a Japanese paper sheet. *3* The dispersion can be applied several times, waiting a few seconds between each application, until it is "rejected" by the porous matrix. *4* After the application, it is necessary to put a cellulose poultice, soaked with distilled water, on the treated surface (still over Japanese paper). The poultice must be humid but not dripping. *5–7* The poultice is left until it dries, then it can be removed gently (along with the Japanese paper). (Image by Michele Baglioni)

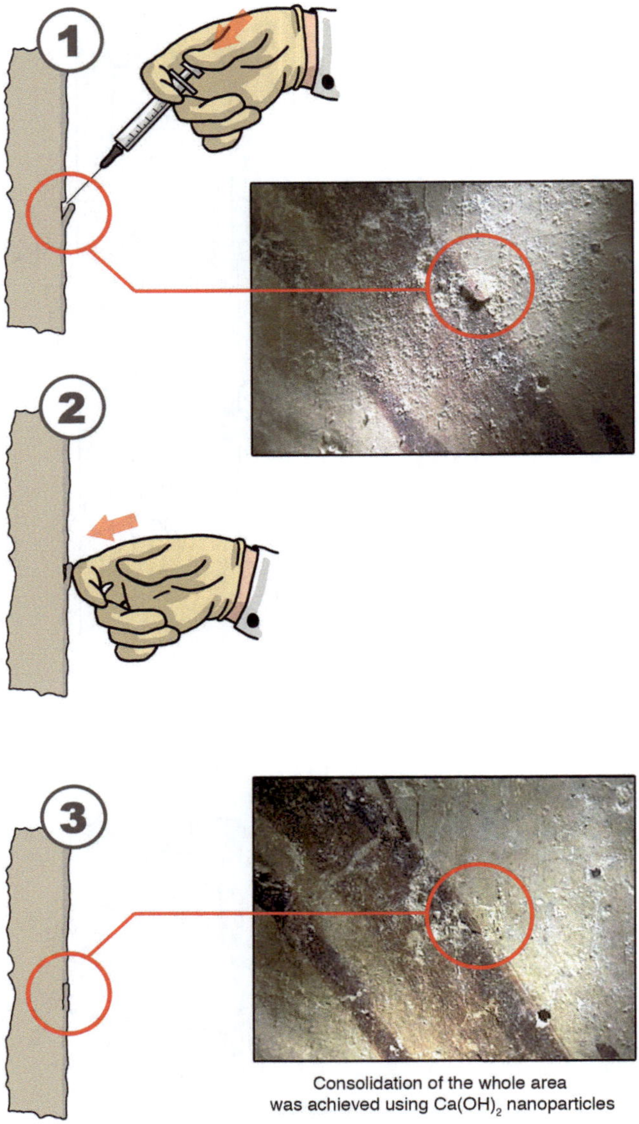

Fig. 2.18 Application of a dispersion of $Ca(OH)_2$ nanoparticles using a syringe to inject the dispersion behind flaking scales and detachments that need to be re-adhered. *1* Injection must be carried out carefully to avoid dripping of the dispersion on the artifact surface. *2* Depending on the surface mechanical resistance, the operator can apply light pressure to facilitate the re-adhesion of the scale, as he/she injects the dispersion. ***3*** The full consolidation of the flaking scale occurs after carbonation of the nanoparticles. Full carbonation might require, indicatively, from 1 to 4 weeks (see text for more details on the carbonation process). Multiple injections might be necessary to achieve consolidation. The consolidation of the whole area showed in the photograph was carried outusing $Ca(OH)_2$ nanoparticles, to consolidate the surface and recover the powdering colors. (Illustrations by Michele Baglioni)

Fig. 2.19 Application by spraying of a dispersion of $Ca(OH)_2$ nanoparticles for the consolidation of a substrate. *1* The application of the nanoparticles dispersion is carried out using a sprayer (see text for further info). The dispersion can be applied several times, waiting a few seconds between each application, until it is "rejected" by the porous matrix. *2* After the application, it is necessary to put a cellulose poultice, soaked with distilled water, on the treated surface (still over Japanese paper). The poultice must be humid but not dripping. *3–5* The poultice is left until it dries, then it can be removed gently (along with the Japanese paper). (Image by Michele Baglioni)

References

Ambrosi M, Dei L, Giorgi R et al (2001a) Colloidal particles of $Ca(OH)_2$: properties and applications to restoration of frescoes. Langmuir 17:4251–4255

Ambrosi M, Dei L, Giorgi R et al (2001b) Stable dispersions of $Ca(OH)_2$ in aliphatic alcohols: properties and application in cultural heritage conservation. Progr Colloid Polym Sci 118:68–72

Armelao L, Bassan A, Bertoncello R et al (2000) Silica glass interaction with calcium hydroxide: a surface chemistry approach. J Cult Herit 1:375–384

Arnold A (1987) The conservation of wall paintings, Symposium held in London, July 1987. The Courtauld Institute of Art and The Getty Conservation Institute Publishers, London, p 103

Ashurst J (1990) Mortars for stone buildings. In: Ashurst A, Dimes FG (eds) Conservation of building and decorative arts, vol 2. Butterwoths, London, p 78

Baglioni P, Giorgi R (2006) Soft and hard nanomaterials for restoration and conservation of cultural heritage. Soft Matter 2:293–303

Baglioni P, Giorgi R (2013) Inorganic nanomaterials for the consolidation of wall paintings and stones. In: Baglioni P, Chelazzi D (eds) Nanoscience for the conservation of works of art. RSC Nanoscience & Nanotechnology No. 28. The Royal Society of Chemistry, Cambridge, p 350

Baglioni P, Dei L, Ferroni E et al (1996) Sospensioni stabili di idrossido di calcio, (Consorzio Interuniversitario per lo Sviluppo dei Sistemi a Grande Interfase, C.S.G.I., Italy). Italian patent FI/96/A/000255, 31 Oct 1996

Baglioni P, Cesari C, Dei L et al (2001) Stucco forte veneziano con Tecniche della Tradizione: proprietà chimico-fisiche, degrado in ambiente lagunare e conservazione mediante nanofasi cristalline di $Ca(OH)_2$. In Atti del XVII Convegno Internazionale Scienza e Beni Culturali -Lo Stucco- Cultura, Tecnologia, Conoscenza, Bressanone, 2001, p 289

Baglioni P, Dei L, Giorgi R et al (2002) Basic suspensions: their preparation and use in processes for paper deacidification (Consorzio Interuniversitario per lo Sviluppo dei Sistemi a Grande Interfase, C.S.G.I., Italy). International patent pending PCT/EP02/00319, 15 Jan 2002

Baglioni P, Carrasco Vargas R, Chelazzi D et al (2006) The Maya site of Calakmul: in situ preservation of wall paintings and limestone using nanotechnology. In: Saunders D, Townsend JH (eds) Proceedings of the IIC Congress 2006, Munich—the object in context: crossing conservation boundaries, James and James, London, p 162

Baglioni M, Berti D, Teixeira J et al (2012) Nanostructured surfactant-based systems for the removal of polymers from wall paintings: a small-angle neutron scattering study. Langmuir 28:15193–15202

Baglioni P, Chelazzi D, Giorgi R et al (2013) Commercial $Ca(OH)_2$ nanoparticles for the consolidation of immovable works of art. Appl Phys A. doi:10.1007/s00339-013-7942-6

Baglioni P, Carrasco Vargas R, Cordeiro Baqueiro M et al (2014) Maya mural paintings in Calakmul: pictorial technique and conservation. In Sgamellotti A, Brunetti BG, Miliani C (eds) Science and art, the painted surface. The Royal Society of Chemistry, London, p 68

Beruto DT, Botter R (2000) Liquid-like H_2O adsorption layers to catalyze the $Ca(OH)_2/CO_2$ solid–gas reaction and to form a non-protective solid product layer at 20 °C. J Eur Ceram Soc 20:497–503

Bonell DGR (1934) The properties of calcium hydroxide I. J Soc Chem Ind 53:279–282

Carretti E, Dei L (2004) Physicochemical characterization of acrylic polymeric resins coating porous materials of artistic interest. Prog Org Coat 49:282–289

Carretti E, Giorgi R, Berti D et al (2007) Oil-in-water nanocontainers as low environmental impact cleaning tools for works of art: two case studies. Langmuir 23:6396–6403

Chelazzi D, Poggi G, Jaidar Y et al (2013) Hydroxide nanoparticles for cultural heritage: consolidation and protection of wall paintings and carbonate materials. J Colloid Interf Sci 392:42–49

Chiantore O, Lazzari M (2001) Photo-oxidative stability of Paraloid protective polymers. Polymer 42:17–27

Ciliberto E, Condorelli GG, La Delfa S et al (2008) Nanoparticles of $Sr(OH)_2$: synthesis in homogeneous phase at low temperature and application for cultural heritage artefacts. Appl Phys A 92:137–141

Croveri P, Dei L, Giorgi R et al (2004) Consolidation of Globigerina limestone (Malta) by means of inorganic treatments: preliminary results. In: Kwiatkowski D, Lofvendahl R (eds) Proceedings of the 10th international congress on deterioration and conservation of stone, vol 1, Stockholm, p 463

D'Armada P, Hirst E (2012) Nano-lime for consolidation of plaster and stone. J Architect Conserv 1:63–80

Dahene A, Herm C (2013) Calcium hydroxide nanosols for the consolidation of porous building materials—results from EU-STONECORE. Herit Sci 1:11–20

Daniele V, Taglieri G (2010) Nanolime suspensions applied on natural lithotypes: the influence of concentration and residual water content on carbonation process and on treatment effectiveness. J Cult Herit 11:102–106

Daniele V, Taglieri G, Quaresima R (2008) The nanolimes in cultural heritage conservation: characterisation and analysis of the carbonation process. J Cult Herit 9:294–301

Daniele V, Taglieri G, Gregori A (2013) Synthesis of $Ca(OH)_2$ nanoparticles aqueous suspensions and interaction with silica fume. Adv Mat Res 629:482–487

De Zorzi C, Favaro M, Tomasin P et al (2009) Consolidation of stone and wall paintings by calcium metallorganic precursors of calcite. In: Proceedings of the 4th international congress on "science and technology for the safeguard of Cultural Heritage in the Mediterranean basin", vol II, Cairo, Egypt, 6–8 Dec 2009, p 433

Dei L, Salvadori B (2006) Nanotechnology in cultural heritage conservation: nanometric slaked lime saves architectonic and artistic surfaces from decay. J Cult Herit 7:110–115

Dei L, Favaro M, Giorgi R et al (2001) Gli stucchi della Cappella della Madonna di Lourdes nella chiesa di S. Giovanni Evangelista in Venezia: caratterizzazione del degrado e preconsolidamento mediante dispersioni alcoliche di calce. In Atti del XVII Convegno Internazionale Scienza e Beni Culturali -Lo Stucco- Cultura, Tecnologia, Conoscenza, Bressanone, 2001, p 557

Dei L, Radicati B, Salvadori B (2005) Sperimentazione di un consolidante a base di idrossido di calico nanofasico sugli affreschi della Cappella del Podestà al Museo del Bargello di Firenze: aspetti chimico-fisici e prove di colore. In: Sulle Pitture Murali, Arcadia Ricerche, Bressanone, p 99

Dei L, Bandini F, Felici A et al (2007) Il Consolidamento degli apparati architettonici e decorative. Arcadia Ricerche, Bressanone, p 217

Elert K, Rodriguez-Navarro C, Sebastian E et al (2002) Lime mortarts for the conservation of historic buildings. Stud Conserv 47:62–75

Espinosa A (1987) In: Hodges HWM (ed) Proceedings of in situ archaeological conservation, INAH, Mexico City, 6–13 April 1986, 1987, p 84

Favaro M, Mendichi R, Ossola F et al (2006) Evaluation of polymers for conservation treatments of outdoor exposed stone monuments. Part I: photo-oxidative weathering. Polym Degrad Stabil 91:3083–3096

Favaro M, Tomasin P, Ossola F et al (2008) A novel approach to consolidation of historical limestone: the calcium alkoxides. Appl Organomet Chem 22:698–704

Ferreira Pinto AP, Delgado Rodrigues J (2008) Stone consolidation: the role of treatment procedures. J Cult Herit 9:38–53

Ferroni E, Baglioni P (1986) Scientific methodologies applied to works of art. In: Parrini PL (ed) Proceedings of the symposium, Florence, Italy, 2–5 May 1984, Montedison progetto cultura, Milan, 1986, p 108

Ferroni E, Malaguzzi V, Rovida V (1969) In: Proceedings of the ICOM-CC plenary meeting, the international council of museums-committee for conservation, Amsterdam, Netherlands, 1969

Fratini E, Miles GP, Giorgi R et al (2007) Competitive surface adsorption of solvent molecules and compactness of agglomeration in calcium hydroxide nanoparticles. Langmuir 23:2330–2338

Giorgi R, Dei L, Baglioni P (2000) A new method for consolidating wall paintings based on dispersions of lime in alcohol. Stud Conserv 45:154–161

Giorgi R, Dei L, Ceccato M et al (2002) Nanotechnologies for conservation of cultural heritage: paper and canvas deacidification. Langmuir 18:8198–8203
Giorgi R, Chelazzi D, Baglioni P (2005a) Nanoparticles of calcium hydroxide for wood conservation. The deacidification of the Vasa warship. Langmuir 21:10743–10748
Giorgi R, Bozzi C, Dei L et al (2005b) Nanoparticles of $Mg(OH)_2$: synthesis and application to paper conservation. Langmuir 21:8495–8501
Giorgi R, Baglioni M, Berti D et al (2010a) New methodologies for the conservation of cultural heritage: micellar solutions, microemulsions, and hydroxide nanoparticles. Acc Chem Res 43:695–704
Giorgi R, Ambrosi M, Toccafondi N et al (2010b) Nanoparticles for cultural heritage conservation: calcium and barium hydroxide nanoparticles for wall painting consolidation. Chem Eur J 16:9374–9382
Gomez-Villalba LS, Lopez-Arce P, Alvarez de Buergo M et al (2011) Structural stability of a colloidal solution of $Ca(OH)_2$ nanocrystals exposed to high relative humidity conditions. Appl Phys A 104:1249–1254
Gomez-Villalba LS, Lopez-Arce P, Fort R (2012) Nucleation of $CaCO_3$ polymorphs from a colloidal alcoholic solution of $Ca(OH)_2$ nanocrystals exposed to low humidity conditions. Appl Phys A 106:213–217
Grassi S, Carretti E, Pecorelli P et al (2007) The conservation of the Vecchietta's wall paintings in the Old Sacristy of Santa Maria della Scala in Siena: the use of nanotechnological cleaning agents. J Cult Herit 8:119–125
Gregory J, Tadros TF (1987) Flocculation by polymers and polyelectrolytes. In: Tadros TF (ed) Solid/liquid dispersions. Academic, New York, p 163
Hamada S, Matijevic E (1982) Formation of monodispersed colloidal pseudocubic haematite particles in ethanol + water solutions. J Chem Soc Faraday Trans 1:2147–2156
Kim EK, Won J, DO J et al (2009) Effects of silica nanoparticle and GPTMS addition on TEOS-based stone consolidants. J Cult Herit 10:214–221
Kingery WD, Vandiver PB, Prickett M (1988) The beginnings of pyrotechnology, part II: production and use of lime and gypsum plaster in the pre-pottery Neolothic Near East. J Field Archaeol 15:219–244
Lazzari M, Chiantore O (2000) Thermal-ageing of Paraloid acrylic protective polymers. Polymer 41:6447–6455
Licchelli M, Malagodi M, Weththimuni M et al (2014) Nanoparticles for conservation of bio-calcarenite stone. Appl Phys A 114:673–683
López T, Dávila VM (2002) In: Kars H, Burke E (eds) Proceedings of the 33rd international symposium on archaeometry, 22–26 April 2002, Geoarchaeological and Bioarchaeological Studies
Lopez-Arce P, Gomez-Villalba LS, Pinho L et al (2010) Influence of porosity and relative humidity on consolidation of dolostone with calcium hydroxide nanoparticles: effectiveness assessment with non-destructive techniques. Mater Charact 61:168–184
Lopez-Arce P, Gomez-Villalba LS, Martinez-Ramirez S et al (2011) Influence of relative humidity on the carbonation of calcium hydroxide nanoparticles and the formation of calcium carbonate polymorphs. Powder Technol 205:263–269
Lopez-Arce P, Zornoza-Indart A, Gomez-Villalba LS et al (2013) Short- and longer-term consolidation effects of portlandite ($Ca(OH)_2$) nanoparticles in carbonate stones. J Mater Civ Eng 25:1655–1665
Maryniak-Piaszczynski E, Egloffstein P, Ziegenbalg G et al (2008) The portal in Tholey—Unconventional method for the preservation of scaling and shelled sandstone (Rotliegend-Sandstone). In: Proceedings of the 11th international congress on deterioration and conservation of stone, Torum, Poland, 2008, p 1247
Matijevic E, Cimas S (1987) Formation of uniform colloidal iron (III) oxides in ethylene glycol–water solutions. Colloid Polym Sci 265:155–163
Matteini M (1991) In: Cather S (ed) Proceedings of the symposium organized by the Courtauld Institute of Art and the Getty Conservation Institute, London, 13–16 July 1987, Getty Conservation Institute, Marina del Rey, 1991, p 137

Matteini M, Moles A (1984) In: Proceedings of the ICOM committee for conservation 7th triennial meeting: Copenhagen, 10–14 Sept 1984: preprints, Paris, 1984, p 84.15.15–84.15.19
Miliani C, Velo-Simpson ML, Scherer GW (2007) Particle-modified consolidants: a study on the effect of particles on sol–gel properties and consolidation effectiveness. J Cult Herit 8:1–6
Mosquera MJ, de los Santos DM, Montes A et al (2008) New nanomaterials for consolidating stone. Langmuir 24:2772–2778
Nanni A, Dei L (2003) $Ca(OH)_2$ nanoparticles from w/o microemulsions. Langmuir 19:933–938
Natali I, Saladino ML, Andriulo F et al (2014) Consolidation and protection by nanolime: recent advances for the consolidation of the graffiti, *Carceri dello Steri* Palermo and of the 18th century lunettes, SS. Giuda e Simone Cloister, Corniola (Empoli). J Cult Herit 15:151–158
Ossola F, Tomasin P, De Zorzi C et al (2012) New calcium alkoxides for consolidation of carbonate rocks. Influence of precursors' characteristics on morphology, crystalline phase and consolidation effects. New J Chem 36:2618–2624
Peroni S, Tersigni C, Torraca G et al (1982) Lime based mortars for the repair of ancient masonry and possible substitutes. In: Proceedings of the ICCROM symposium on Mortars cements and grouts used in the conservation of historic buildings, ICCROM, Rome, 1982, p 63
Piqué F, Dei L, Ferroni E (1992) Physicochemical aspects of the deliquescence of calcium nitrate and its implications for wall painting conservation. Stud Conserv 37:217–227
Poggi G, Toccafondi N, Melita LN et al (2014) Calcium hydroxide nanoparticles for the conservation of cultural heritage: new formulations for the deacidification of cellulose-based artifacts. Appl Phys A 114:685–693
Riederer J (1984) Committee for Conservation, 7th Triennial meeting, Copenhagen. ICOM, Paris, p 21
Rodriguez-Navarro C, Hansen E, Ginell WS (1998) Calcium hydroxide crystal evolution upon aging of lime putty. J Am Ceram Soc 81:3032–3034
Rodriguez-Navarro C, Ruiz-Agudo E, Ortega-Huertas M et al (2005) Nanostructure and irreversible colloidal behavior of $Ca(OH)_2$: implications in Cultural Heritage conservation. Langmuir 21:10948–10957
Rodriguez-Navarro C, Suzuki A, Ruiz-Agudo E (2013) Alcohol dispersions of calcium hydroxide nanoparticles for stone conservation. Langmuir 29:11457–11470
Salvadori B, Dei L (2001) Synthesis of $Ca(OH)_2$ nanoparticles from diols. Langmuir 17:2371–2374
Son S, Won J, Kim JJ et al (2009) Organic–inorganic hybrid compounds containing polyhedral oligomeric silsequioxane for conservation of stone heritage. ACS Appl Mater Interf 2:393–401
Vincent B, Bijsterbosch BH, Lyklema J (1971) Competitive adsorption of ions and neutral molecules in the stern layer on silver iodide and its effects on colloid stability. J Colloid Interf Sci 37:171–178
Volpe R, Taglieri G, Daniele V et al (2013) A process for the synthesis of $Ca(OH)_2$ nanoparticles by means of ionic exchange resins. PCT/IB2013/056195, 29 July 2013
Yura K, Fredrikson KC, Matijevic E (1990) Preparation and properties of uniform colloidal indium compounds of different morphologies. Colloids Surf A 50:281–293
Ziegenbalg G (2008) Colloidal calcium hydroxide: a new material for consolidation and conservation of carbonatic stones. In: Proceedings of the 11th international congress on deterioration and conservation of stone, Torun, Poland, 2008, pp 1109

Chapter 3
Cleaning of Wall Paintings and Stones

Abstract The surface of artifacts interacts with the environment, leading to aging and degradation. Soiling is one of the main factors that contribute to the alteration of artistic and historical surfaces, and is often associated with both chemical and mechanical degradation. The effects of these processes on porous inorganic substrates such as wall paintings and stone are usually strongly enhanced if organic coatings (e.g. acrylic and vinyl-based polymers), applied in previous restoration treatments, are present. In fact coatings alter the physico–chemical properties of the artifacts (e.g. water vapor permeability), enhancing the degradation processes promoted by salts and pollution agents, ultimately the detachment and loss of the painted layer. In the recent years innovative formulations based on nanostructured fluids have been developed and successfully used for the removal of both grime/dirt and detrimental synthetic or natural coatings. This chapter focuses on the formulation and application of surfactant-based complex fluids, such as microemulsions and micellar solutions, for the removal of dirt and coatings from the surface of monuments and wall-paintings. Due to the low content of organic solvents, these systems have a very low eco-toxicological impact as compared to pure solvents or solvent blends; moreover, the nanostructure of the cleaning fluids grants the effective and selective removal of undesired layers.

3.1 Background

In the framework of cultural heritage preservation, compatibility is defined as the use of materials whose physico–chemical properties are similar to (or possibly the same as) those of the works of art to be treated. This fundamental concept is gaining importance in modern restoration practice. The application of physico–chemically incompatible materials (i.e. organic polymers) to porous inorganic immovable works of art, such as wall-painting and stone, is presently often discouraged, however in the last four decades many masterpieces have been treated with protective/consolidant polymer coatings and they presently exhibit consistent degradation.

The application of synthetic polymers on mural paintings was widely diffused due to several factors that made these products appealing in the short term: first,

P. Baglioni et al., *Nanotechnologies in the Conservation of Cultural Heritage*,
DOI 10.1007/978-94-017-9303-2_3

the high adhesive power of synthetic formulations provided a quick solution to the issue of flaking parts and detachment of painted layer from the surface of wall paintings. These coatings also provide appealing short-term effects such as gloss and intensity to colors. The introduction of synthetic polymers for the preservation of cultural heritage was also favored by the misbelief of their permanent resistance to aging and degradation. Moreover, the use of these materials was prompted by the recommendations of scientists worldwide (Philippot and Mora 1969; Mora et al. 1984; Horie 1987). Finally, both the low costs of these materials and their easy application onto artifacts contributed to their extensive use.

In fact, conservators and scientists initially considered these products as reversible materials, i.e. easily removable by using the same solvents used during their application. However, practical experience has shown that in some conditions these materials favor the degradation of the treated surfaces because the presence of polymers results in strong alteration of the main physicochemical properties of the surface, namely a drastic change of water capillarity, water vapor permeability, and surface wettability (Giorgi et al. 2010). Polymeric coatings applied over the surface block the porous structure of the murals and prevent water from evaporating, promoting thus salt crystallization in the inner layers (Giorgi et al. 2010; Magaloni Kerpel 2011). Under these conditions, salt crystals exert pressure on the external layers as they migrate towards the open surface, leading to cracks and mechanical stress; as a result, the substrate is disaggregated and in the best case the flaking parts remain glued to the polymeric film, which eventually detaches from the surface, causing irreversible damages to the artifact (see Fig. 3.1a,b). These processes can lead in the medium or long term (e.g. from less than 10 to 50 years depending on the environmental conditions) to complete disruption and detachment of the painted layer. In the late 1980s, conservators reported this issue while working on the preservation of wall paintings and plasters in tropical sites, where polymeric coatings had caused the disruption of the surface layers of the works of art (Cedillo 1991; López Ortega 1999).

Furthermore, due to natural aging even synthetic polymers can undergo degradation as for instance molecular weight changes due to cross-linking reactions and chain scissions. The main consequence of cross-linking is that the polymers are no longer soluble in the same solvents used to apply them, hampering their removal and making their application irreversible (Lazzari and Chiantore 2000; Chiantore and Lazzari 2001; Favaro et al. 2006; Giorgi et al. 2010). Finally, aging can also produce discoloration (yellowing) of the coatings, resulting in aesthetical alteration that hinders the readability of the original works of art. Currently, the issue of degradation induced by incompatible coatings on wall paintings has reached its peak and the cultural heritage of several civilizations is at risk, making the removal of acrylic and vinyl co-polymers a central topic.

Because of the degradation processes described above, the original pictorial layers lying beneath the coatings can be degraded and fragile. Therefore, the cleaning intervention is often followed by, or coupled with, the strengthening of the weakened layers (the use of nanotechnology for consolidation of mural paintings and stone has been detailed in Chap. 2). When pictorial layers are particularly fragile, a pre-consolidation step can be carried out before the cleaning intervention. Cleaning and consolidation cycles can also be alternated on fragile painted surfaces to ensure that the loss of original artistic materials is minimized. It is important to

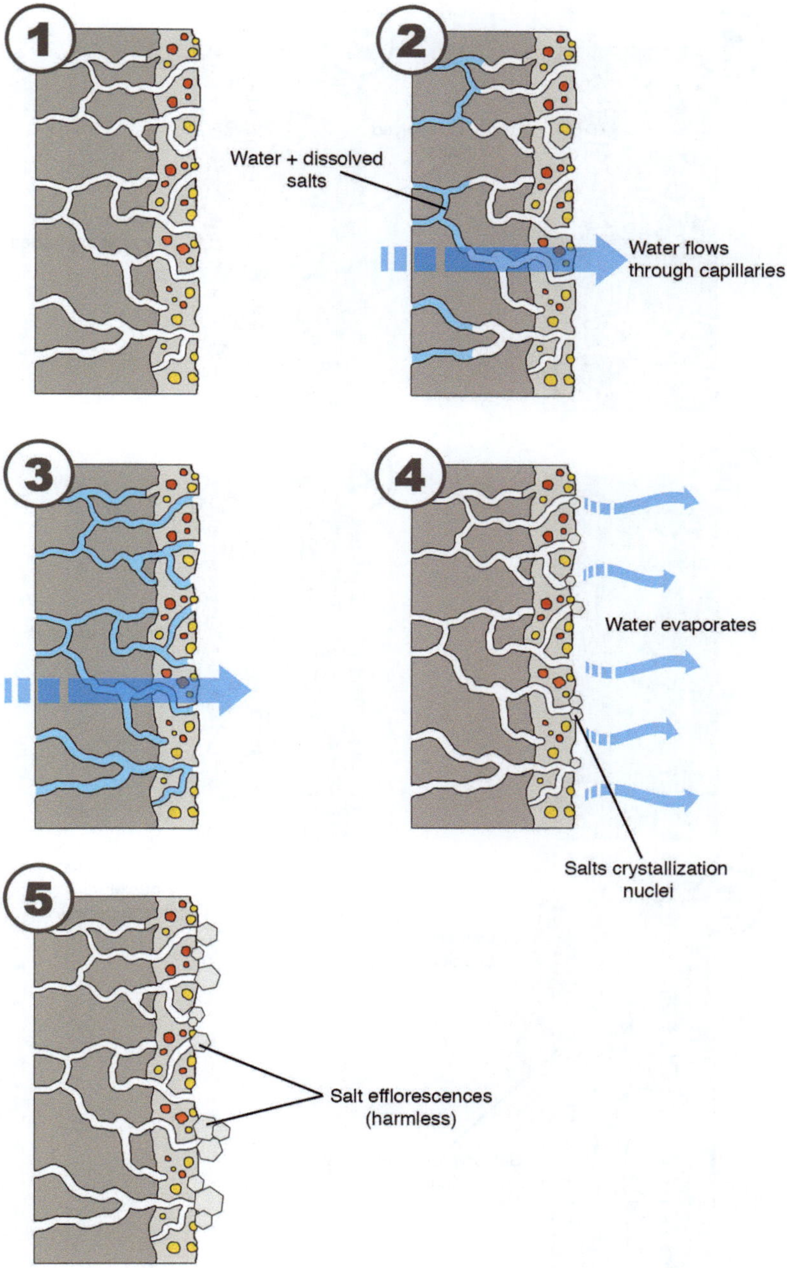

Fig. 3.1a Schematic representation of the formation of salt efflorescence on a wall painting that was not treated with polymeric coatings. (Image by Michele Baglioni)

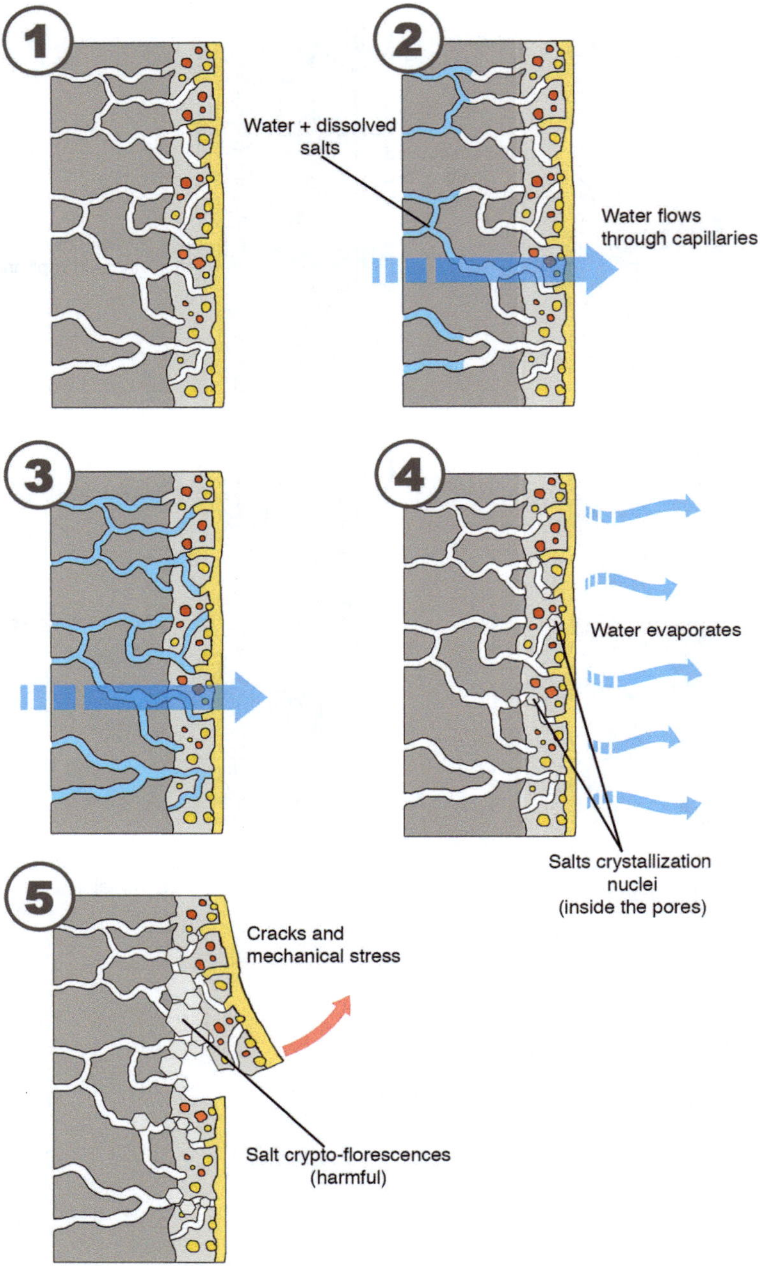

Fig. 3.1b Schematic representation of the degradation effects due to the presence of polymer coatings on the salt crystallization process beneath the painted layer. 1-5 Polymeric coatings applied over the surface block the porous structure of the murals, promoting salt crystallization in the inner layers. Under these conditions, salt crystals exert pressure on the external layers as they migrate towards the open surface, leading to cracks and mechanical stress. (Image by Michele Baglioni)

notice that leaving the aged adhesive/coating on the artistic surface might lead in any case to the partial or complete loss of the painted layers, as described above.

As a general criterion, cleaning systems must address three fundamental targets: (1) selectivity and efficiency in removing soil or coatings, (2) full physico–chemical compatibility with the original artistic substrate, and (3) safety for the operator and the environment.

The classic approach to cleaning is based on the use of solvents, either neat or blended. In this chapter, the conceptual difference in the use of advanced cleaning systems such as micellar solutions and microemulsions as compared to solvents will be highlighted.

The effectiveness of cleaning with solvents depends on their ability to dissolve a specific material while leaving other materials unaffected. To that end, the choice of a solvent requires knowledge of several parameters, such as the evaporation rate, solution viscosity, and both environmental and health concerns. The selection of the right solvents or solvent blends is a fine art, based on knowledge, experience and intuition guided by rules of thumb such as "like dissolves like" and various definitions of solvent "strength" (Burke 1984).

Some tools are nowadays available to conservators and probably the Teas charts are one of the most used. These were developed to illustrate in a concise way the fractional solubility parameters that Teas defined starting from the Hansen solubility parameters, which derived from earlier work of Hildebrand on polymer solubility. Hildebrand established that the cohesive energy density (CED) i.e. the strength of the overall interactions in the solute (the substance dissolved in a solution), must be matched for solubility to occur. Hansen improved on Hildebrand's solubility parameter, δ, by further dividing the CED value into components based on specific intermolecular forces, i.e. dipolar interactions (polarity), hydrogen bonding interactions, and dispersion (London) forces. Therefore, Hansen parameters consist of three values that represent the ability of a material to give these interactions. The sum of these values gives the Hildebrand parameter.

Teas presented these different contributions as fractional parameters, where each contribution is related to the whole cohesive energy density. Therefore, for each solvent the sum of the three Teas parameters is always 1.

The triangular Teas chart, where each axis represents one Teas parameter, allows the univocal determination of the position of one solvent (or a single blend of solvents) in the diagram and verifies its correspondence to the solubility area of the different class of substances that need to be removed (oil, proteins, synthetic adhesives etc.). In practice, solvents in close proximity on the Teas chart are expected to be mutually miscible, and solvents in proximity to the swelling area of a substance (adhesive, coating, etc.), are expected to swell or dissolve such substance. The Teas chart is thus a handy tool that allows prediction of the solvents to be used for the solubilization of soiling materials or detrimental coatings, based on its molecular structure.

However, in the framework of cleaning works of art one of the main concerns about the direct use of solvents is represented by the physical mechanism involved in the process. In fact, the solubilization of soil or coatings involves that dissolved matter might migrate through capillarity within the porous structure of the artifact. Upon evaporation of the solvent, the solubilized materials re-deposit within the pores of the artifact, rather than being removed (see Fig. 3.2). This obviously

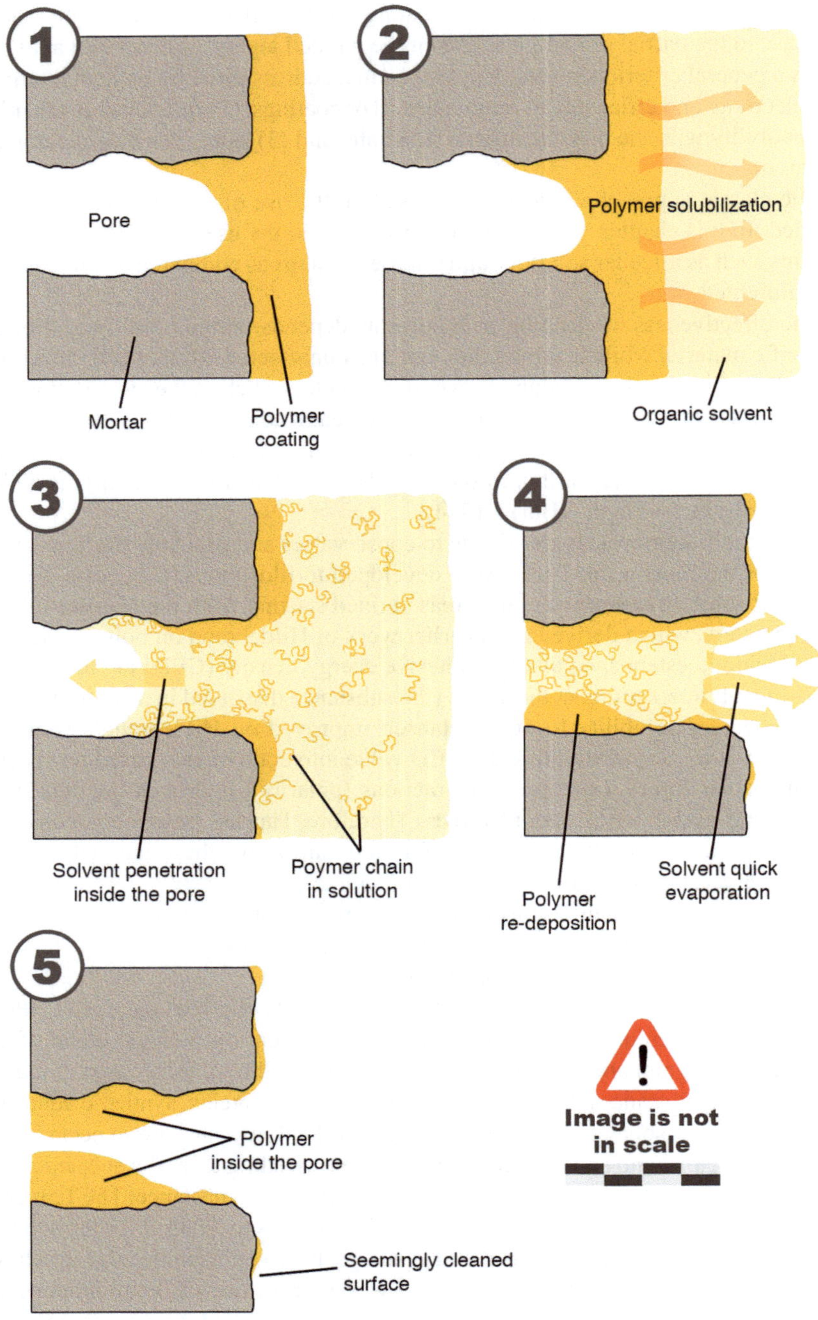

Fig. 3.2 Solubilization of a polymer coating (lying on a mortar) (*1*,*2*), and re-deposition of dissolved matter within the pores of the mortar (*3*–*5*), as a consequence of using an organic solvent. (Image by Michele Baglioni)

depresses the efficacy of the cleaning action, which is usually acceptable only because the visual appearance of the artistic surface is appealing, while the physico–chemical properties of the porous substrate may remain altered by the presence of the re-deposited soil/coating. Highly sorbent materials (e.g. cellulose pulp poultices) can be used to "suck" and extract the dissolved matter through capillarity, removing it from the artifact matrix. However, the high evaporation rate of the solvents typically used for cleaning works of art favors the re-deposition of the solubilized materials within the pores.

Finally, another important concern is represented by the toxicity of pure solvents and solvent blends.

For these reasons, it is of paramount importance to explore different ways to remove soil/grime and detrimental coatings in order to achieve long-lasting results while minimizing health and environmental impact. In most cases water is fully respectful of the physico–chemical characteristics of mural paintings. Thus, the ideal cleaning agent would be water provided with the ability to interact with hydrophobic materials and remove them. Obviously, pure water has no such power. However, the use of water-based amphiphilic systems, such as micellar solutions and oil-in-water microemulsions, allows reaching this challenging task. Indeed, it is possible to generate in the water phase stable structures or aggregates that are able to solubilize or swell hydrophobic materials.

The use of water-based systems offers important advantages. First it may inhibit the re-deposition of the solubilized soil, grime or coatings. In fact, during the application the surface of the artifact pores is covered by a layer of water that prevents re-adhesion of the detached hydrophobic materials (grime, polymers), which are thus effectively removed or extracted by the sorbent (poultices or gels). This process is favored by the fact that water exhibits a lower evaporation rate than that of the organic solvents commonly used for restoration. Another fundamental aspect is that water-based systems such as microemulsions and micellar solutions have low organic content (roughly from 1 to 25 % w/w, including both solvents and surfactants); therefore the impact on the health of the operator and on environment is minimized.

The fine description of the phase behavior, the properties and the structure of water-based amphiphilic systems is beyond the scopes of this book, and the interested reader can refer to a vast amount of literature on this subject. See for instance Rance 1977; Evans and Wennerström 1994; Azemar 1997; Lange 1999; Holmberg et al. 2002; Goodwin 2009; Stubenrauch 2009. In the following paragraphs we provide a fundamental description of the main features that are relevant in the framework of artifacts cleaning.

A micellar solution is a 'dispersion' of aggregated amphiphilic molecules (surfactants); the aggregates typically show a spherical shape and are called 'micelles'. Micelles form when the concentration of surfactants exceeds a threshold value called the critical micellar concentration (CMC); under these conditions micelles are still in equilibrium with free, non-aggregated amphiphiles. The micelles shape depends mainly on the chemical structure of the surfactant; the spherical shape is the most common, but also cylindrical, oblate or more complex structures can be obtained.

The presence of surfactants above the CMC in a blend of two immiscible liquids leads, upon mechanical stirring, to the formation of an emulsion. These systems, which usually show a cloudy appearance, are kinetically stable systems, i.e. upon time the two immiscible phases tend to separate. The emulsion is described with the term "oil-in-water" (o/w) or "water-in-oil" (w/o), depending on which liquid is the continuous phase (dispersing liquid) and which the dispersed phase (the dispersed liquid); in o/w emulsions the hydrophobic liquid (generically indicated as "oil") is dispersed as droplets in the hydrophilic liquid (water). According to the Bancroft rule, when comparable amounts of liquids are used, the solvent where the surfactant is more soluble becomes the dispersing medium.

The droplet size is one of the fundamental differences between "emulsions" and "microemulsions". In standard emulsions, the diameter of droplets ranges from 200 nm up to few microns (Estoe 2005), while microemulsions exhibit droplets of 5–150 nm. The reduced droplet dimension accounts for two important characteristics of microemulsions: (i) microemulsions are optically transparent due to the micelles size; (ii) the nanodroplets develop a huge exchange surface area (see also Chap. 1, Fig. 1.1), which strongly enhances the interaction of these systems with any undesired layers that need to be removed (grime, detrimental adhesives, etc.). As a consequence, the cleaning effectiveness of microemulsions is kinetically enhanced with respect to standard emulsions.

More specifically, a microemulsion is defined as "*a system of water, oil and amphiphile, which is a single optically isotropic and thermodynamically stable liquid solution*" (Danielsson and Lindman 1981). Similarly to emulsions, if the dispersed phase is oil and the continuous phase is water, microemulsions are defined oil-in-water (o/w); if the reverse occurs, microemulsions are classified as water-in-oil (w/o) (see Fig. 3.3).

In microemulsions, the surface tension (γ) at the interface between oil and water is extremely low (tending to $\gamma = 0$) due to the presence of surfactants and co-surfactants (small molecules that act in addition to surfactant, further reducing surface tension). In such conditions, spontaneous emulsification of the water and oil phases occurs, leading to a thermodynamically stable system (Holmberg et al. 2002). This property radically differentiates microemulsions from emulsions, as the latter are only temporarily stable. The formation of microemulsions is a process that strongly depends on the temperature (and pressure) and on the concentration of the three fundamental system components, i.e. the two immiscible liquids and (at least) one surfactant. The process is fully represented with the use of a ternary (triangular) phase diagram where each side of the triangle shows the volume fraction of a single component ranging in a given interval. The co-surfactant is represented as single "pseudo-component", where the cosurfactant amount is at a fixed ratio to surfactant. Therefore, in a phase triangle at fixed pressure and temperature, each point represents a possible composition of the three components (see Fig. 3.4). Different regions of the diagrams (separated by boundaries) correspond to different phases of the system (e.g. "o/w" or "w/o"), which overall represent the "phase behavior" of the system at constant temperature and pressure. Moving along one side of the triangle, the volume fraction of the related component changes and, consequently,

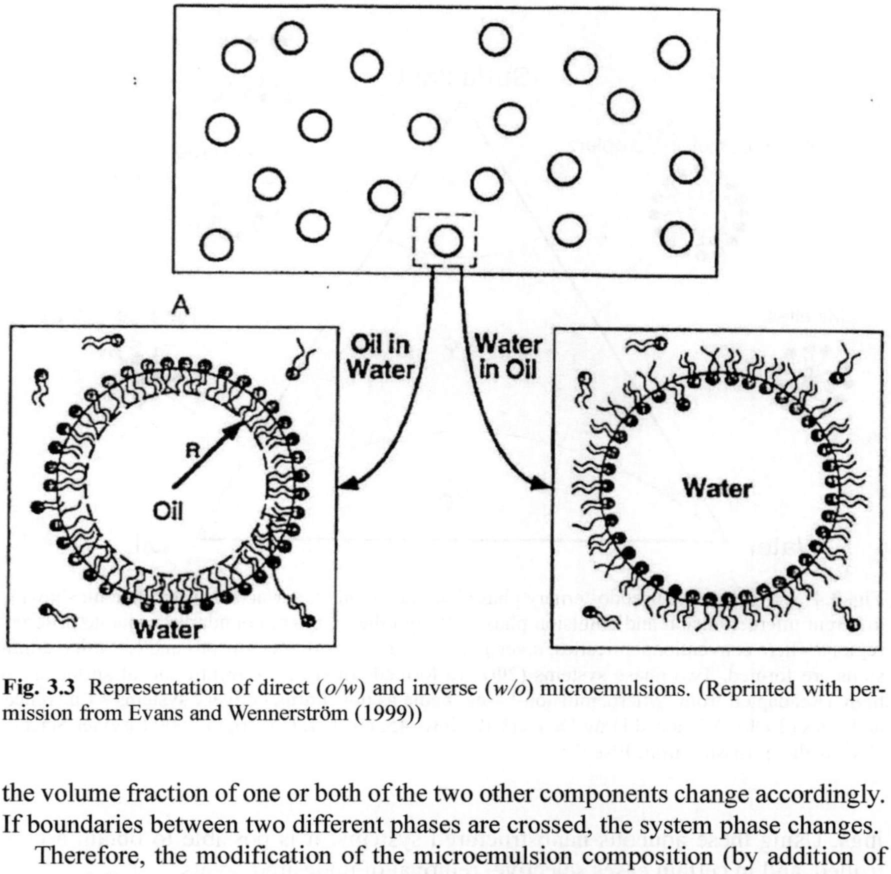

Fig. 3.3 Representation of direct (*o/w*) and inverse (*w/o*) microemulsions. (Reprinted with permission from Evans and Wennerström (1999))

the volume fraction of one or both of the two other components change accordingly. If boundaries between two different phases are crossed, the system phase changes.

Therefore, the modification of the microemulsion composition (by addition of some of the components or because of the evaporation of volatile parts), or the change of temperature (or pressure), involves that the system "is moving" within the phase diagram and, consequently, it might pass through phase boundaries, undergoing phase changes or phase separation, i.e. the disruption of the microemulsion into separated components. Both phase changes and phase separation affect the cleaning effectiveness of microemulsions, which might be significantly altered or reduced. Thus, the experimentation of different compositions and the alteration of the original cleaning systems formulation require deep knowledge of the related diagram phases (see also Sect. 3.4).

Microemulsions are surfactant-based colloidal systems that exhibit a wide number of possible applications, from food science to drug delivery, biotechnology and template directed synthesis of inorganic nanoparticles (Holmberg et al. 2002; Estoe 2005). These systems also represented a significant step forward in conservation science, since their outperforming detergency properties can be used for cleaning purposes. In fact, microemulsions represent the most interesting alternative to the use of neat/blended organic solvents or standard emulsions for the removal of organic materials, such as grime or detrimental synthetic polymers, from wall paint-

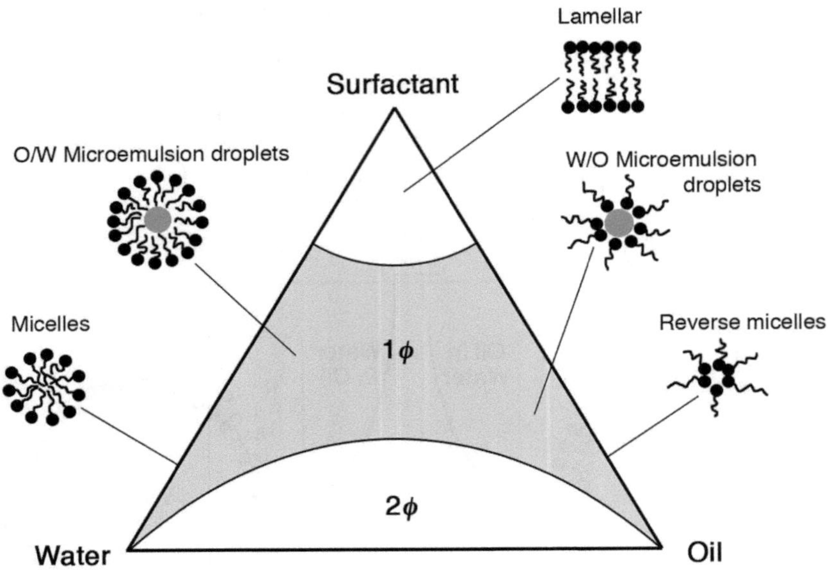

Fig. 3.4 A hypothetical pseudo-ternary phase diagram of an oil/surfactant/water system, showing different microemulsion and emulsion phases. Within the diagram, boundaries separate different regions where conventional micelles, reverse micelles or w/o microemulsions and o/w microemulsions are formed. Two-phase systems (2Φ) are formed when the concentration of surfactant is high. (Readapted from "Microemulsion-based media as novel drug delivery systems", Lawrence, and Rees (2000), Advanced Drug Delivery Reviews 45, 89–121. Copyright 2000 Elsevier Science B.V., with permission from Elsevier)

ings. Using these aqueous nanostructured systems, it is possible to obtain a controlled, and in certain cases selective, removal of undesired layers.

In recent years several surfactant-based cleaning fluids, such as microemulsions and micellar solutions have been designed, prepared and tested for cultural heritage preservation (Carretti et al. 2003, 2007; Giorgi et al. 2010; Baglioni et al. 2012a, b, 2014). To summarize, these systems exhibit the following main advantages with respect to traditional cleaning methods:

- They produce the solubilization and/or the swelling of grime and polymer coatings, causing their detachment from the artistic substrates. The penetration of the detached hydrophobic material within the porous matrix of the substrate (e.g. wall painting) is hindered by the hydrophilic barrier provided by the fluids continuous phase (water). Instead, traditional cleaning with solvents involves the dissolution of coatings and the transport of the dissolved material within the pores of the work of art, with possible detrimental long-term effects.
- They can be effective in removing undesired coatings whose solubility in solvents has been altered and reduced due to aging.
- They are water-based systems with low content of organic solvents, allowing a drastic reduction of the impact on health of end-users (conservators, restorers) as compared to pure solvents.

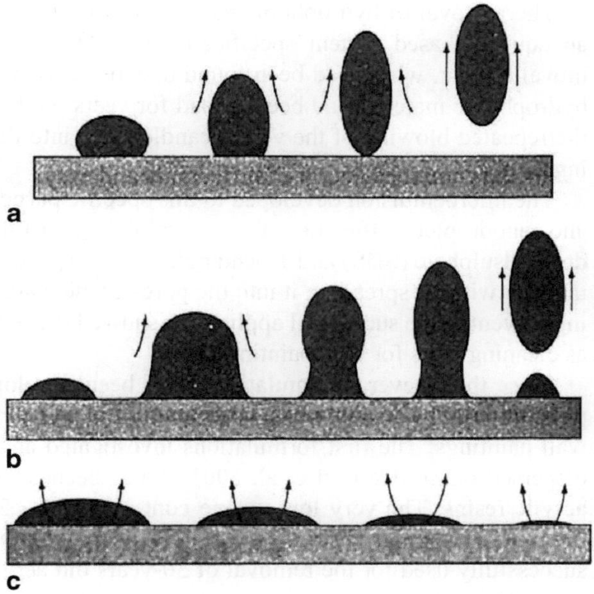

Fig. 3.5 Dynamic representation of the removal mechanisms of oily soils. **a** rolling-up, **b** emulsification, **c** solubilization. (Reprinted with permission from Holmberg et al. (2002) Surfactants and polymers in aqueous solution, 2nd Edn. Copyright © 2002, John Wiley and Sons)

To conclude, it is helpful to shortly recall the main mechanisms involved in the removal of grime and oily soil. As graphically described in Fig. 3.5, these are:

- Rolling-up: This mechanism is relevant when the apolar oily soil is deposited on polar substrates, e.g. oil spots on cotton fabric. In this process the interaction between the substrate and the surfactant is decisive.
- Emulsification: The mechanism involves the interaction of the surfactant aggregates (micelles) with the 'oil' phase. This process is not dependent on the nature of the substrate. Low interfacial tension is needed in this process.
- Solubilization: The oily soil is solubilized into the hydrophobic core of a microemulsion. This mechanism is not dependent from the chemical nature of the surface from which the soil has to be removed. In this process, ultra-low interfacial tension between oil and the surfactant solution is usually needed.
- The removal of polymers might involve more complex processes. Their description is outside the aim of this treatise and the interested reader can refer to the chapter references for further information, for instance see Baglioni et al. (2010, 2012a, b, 2014).

3.2 Formulations

The first application of microemulsions for the conservation of cultural heritage was performed by Ferroni and Baglioni at the end of 1980s, during the restoration of fifteenth century fresco paintings (by Masaccio, Masolino, and Lippi) in the Brancacci Chapel, in Florence (Ferroni et al. 1992; Borgioli et al. 1995, 2013, 2014).

The removal of hydrophobic material from the paintings was performed using an aqueous-based system specifically formulated for the solubilization and removal of wax, which had been found unexpectedly over the painted surface. This hydrophobic material had been spread for years on the painted surface because of the repeated blowing of the votive candles kept into the chapel, close to the paintings.

The microemulsion developed to this specific purpose was composed of dodecane nanodroplets (efficient in the wax solubilization) stabilized in water by sodium dodecylsulphate (SDS) and 1-pentanol. This formulation was effective in removing the wax without spreading it into the pores of the wall, as would have occurred using solvents. The successful application showed the potentiality of microemulsions as cleaning tools for wall paintings.

Since then, several formulations have been developed to remove aged acrylic polymers/co-polymers, which are one of the most used classes of consolidants for wall paintings. The first formulations investigated and tested contained xylene as dispersed phase (Carretti et al. 2001, 2003) because of its high affinity for aged acrylic resins. The very low xylene content (up to $<3\,\%$) decreased dramatically the health and environmental impact of the intervention. These formulations were successfully used for the removal of 30-years old acrylic coatings (applied in past restorations) from fourteenth century wall paintings by Spinello Aretino, in the Guasconi Chapel (San Francesco Church, Arezzo, Italy). The microemulsions were applied using the poultice technique (see Sect. 3.4), and the application led to the full removal of the detrimental coatings both from the surface and from the inner layers of the mural paintings, as confirmed by analysis through FTIR (Fourier transform infrared spectroscopy).

Another class of consolidants largely used in the past (and that nowadays often exhibit degradation) comprises vinyl polymers and vinyl-acrylate co-polymers. One of the first cases where microemulsions were used for the removal of vinyl-based coatings involved the preservation of Renaissance wall paintings decorating the external walls of the Conegliano Cathedral (Treviso, Italy). Because *p*-xylene is not a good solvent for vinyl polymers, different nanostructured fluids were formulated than those used in the previous applications. In fact, a micellar solution, containing SDS, 1-pentanol (as co-surfactant) and propylene carbonate (PC) was successfully used for this task (Carretti et al. 2003), see Fig. 3.6. The complete structural characterization of this system was also performed and the nanostructure, i.e. the size distribution and the shape of the droplets, fully described (Palazzo et al. 2005).

Few years later, a similar system was used to remove acrylic polymers (i.e. Paraloid B72) from mural paintings in the church of Santa Maria della Scala in Siena (Grassi et al. 2007). In 2008 a modified system was developed, which contains SDS, 1-pentanol, PC and ethyl acetate (EA), to remove a vinyl/acrylic co-polymers from Maya wall paintings conserved in the archeological site of Mayapan (Yucatan, Mexico) (Giorgi et al. 2010). This oil-in-water system (named "EAPC") was found to be very effective in the removal of several kinds of polymers. Beside the Mayapan site, EAPC was successfully tested for the removal of acrylic coatings from

Fig. 3.6 Fresco paintings by Pozzoserrato (sixteenth century) on the external walls of the Santa Maria dei Battuti Cathedral in Conegliano (Italy). The *dotted blue line* separates the area where aged vinyl polymers (applied in past restorations) were removed using a micellar solution (*left*) from the non-cleaned region (*right*). (Reprinted with permission from Carretti et al. (2003))

the paintings of the Cholula site (Mexico), and for the removal of silicone-based coatings from the wall decorations of the Annunciation Grotto in Nazareth, Israel (Baglioni et al. 2010, 2012a, b). It must be noticed that in some cases natural aging had altered the detrimental polymer coatings in such a way that traditional solvents were not able to solubilize them, while the water-based nanostructured cleaning systems proved fully effective in removing the aged polymers and recovering the artifact readability (see Fig. 3.7).

Recently, research activity on the development of new microemulsion formulations has been addressed to the use of nonionic surfactants because they aggregate at lower concentration and they are more environment-friendly and biodegradable than ionic ones (such as SDS). Alkyl polyglycosides (APG) and ethoxylated alcohols belong to a class of "green" surfactants that possess interesting properties for a great number of applications, including cleaning of works of art (Luders 2000; Carretti et al. 2007). More recently, fully biodegradable non-ionic surfactants (broad- or narrow-range ethoxylated alcohols, less than 7 %) were used to formulate a cleaning system, based on methyl ethyl ketone, ethyl acetate and butyl acetate (less than 5 % each).

It is fundamental to recall that the preparation of microemulsions and micellar solutions requires some chemical skills and expertise in the field of colloid chemistry. It is not a difficult procedure for well-practiced chemists and does not require special and sophisticated equipment, but a full comprehension of the physical-chemistry involved in each preparation step is mandatory.

The feasible preparation and the cheap chemicals used, make these formulations not expensive, since the larger component is simply distilled water; in fact, large amounts of these systems can be prepared in a single shot if large tanks are available. However, in order to obtain high-quality and high-performing systems, the supervision of the preparation procedure by skilled personnel is highly recommended. This technology for the preservation of Cultural Heritage is protected by

Fig. 3.7 Application of the EAPC nanostructured system on wall paintings from the Annunciation Basilica in Nazareth (Israel). (*Top*) Before restoration. (*Bottom*) After restoration. In the dashed box, an area is highlighted where the polymer coating has been left on the painting as a reference for the evaluation of the cleaning result. (Reprinted with permission from Baglioni et al. (2012b)

the trademark Nanorestore Cleaning® as a warranty of quality of the products and of correspondence to the original formulations; these formulations are not distributed on the market yet, but CSGI (Consorzio Interuniversitario per lo Sviluppo dei Sistemi a Grande Interfase, Center for Colloid and Surface Science)—University of Florence is available to provide conservators with materials and technical assistance on the basis of specific agreements.

3.3 Application Field

The removal of polymer coatings using solvents is unadvisable because of several reasons as already reported in the previous sections: the cleaning is scarcely controlled and poorly selective, as the polymer is dissolved by the solvent and spread throughout the inner pores of the substrate. Finally, toxicity of most organic solvents represents an issue that cannot be neglected about the safety of the operators.

Water-based detergent solutions, such as micelllar solutions and microemulsions, allow minimizing these concerns both when these cleaning systems are used in outdoor and indoor conditions, and also on materials with different porosity.

These systems were originally developed for the removal of hydrophobic coating from hydrophilic surfaces, i.e. mural paintings and stones. However, in principle, micellar solutions and microemulsions may have a wider application field, since they can be also used on other hydrophilic materials used in art as stucco, ceramics, cement, glass, bones, metals, wood, textiles, and others. Most of the works of art materials cited above have been treated in the past with hydrophobic coatings, and, therefore, the applicability of this new approach for cleaning based on nanostructured fluids has very large perspectives (Chevalier et al. 2008; Carretti and Giorgi 2013).

Microemulsions and micellar solutions can be used wherever necessary to remove soil/grime and undesired layers (detrimental coatings, aged varnishes, adhesives, etc.) that present a different affinity in terms of hydrophobicity/hydrophilicity with the substrates. The same approach can be followed when the substrate is sensitive to water and the removal of water-soluble soil is required. In such a case, inverted micelles or water-in-oil microemulsions may provide a successful cleaning of the surface.

More recently, the applicability of oil-in-water systems on water sensitive substrates, such as canvas and wood paintings, parchment and paper, has been investigated. This issue is highly relevant since polymer coatings were largely used in the past also on these materials. The use of water as a continuous phase might be invasive due to the high water sorption of the substrates and the influence that the water content has on the mechanical properties of these supports. In fact, swelling and shrinkage due to wetting/drying processes may represent a risk for conservation of these artifacts. However, the application technique based on the use of highly retentive chemical hydrogels (that will be described in detail in Chap. 4), able to load microemulsions and micellar solutions, allows excellent results, as the effective removal of soil/grime or polymers is achieved with no detrimental effects on the works of art substrates (see for instance Pizzorusso et al. 2012).

3.4 Application Procedures

The aim of this section is to provide practical guidelines for the application of microemulsions and micellar solutions on inorganic porous substrates such as wall paintings and stones. In particular, these guidelines are to be used for the application of nanostructured cleaning fluids as designed at CSGI, and registered as Nanorestore Cleaning®. Different formulations might behave differently, so adaptations of the indications here provided should be considered case by case.

The proposed methods are based on the theoretical and practical knowledge gathered in the last decades through both research work and field activity in collaboration with conservators. It is fundamental to recall that conservation issues can vary greatly; therefore the practical guidelines reported here should be considered and applied critically, and not as recipes.

Conservators are encouraged to contact the authors of this book for further guidance and recommendations on the use of these formulations. In particular, the experimentation of different compositions and the alteration of the original cleaning systems require deep knowledge of the related diagram phases; otherwise the system effectiveness might be significantly altered or reduced. The shelf life of the EAPC microemulsion is ca. 2-3 months. After this time the ethyl acetate present in the microemulsion slowly degrades and although the microemulsion can still be used for cleaning, a post-cleaning treatment with $Ca(OH)_2$ nanoparticles (see Sect. 2.4.2) is recommended.

The cleaning of surfaces is an irreversible process, therefore the methodology chosen to remove soil or polymeric coatings must allow full control of all the steps, in order to ensure the complete solubilization/swelling and removal of the undesired materials while avoiding their penetration within the painting pores, so that the physicochemical properties of the mural are fully re-established. On the contrary, the redistribution of solubilized polymers within the painting pores normally occurs after direct solubilization of the coatings with pure or blended solvents.

The best practice for the application of micellar solution/microemulsions requires a simple preparatory work to test the stability of the paintings when liquids are applied on top, and then the use of the poultice technique for the direct application of the nanostructured fluids.

The preliminary stability tests are carried out using the microemusions. The aim is to test the sensitiveness to swelling or solubilization of both pigments and binding media present in the paintings. To this end, small amounts of the fluids can be applied on small areas either directly or soaked in a swab, checking possible negative effects.

O/w microemulsions and aqueous micellar solutions are usually applied with the cellulose-pulp poultices technique (for the application of fluids loaded in gels, see Chap. 4). The application of the microemulsion-loaded poultice (the same applies also for micellar solutions) consists in the following procedure:

Microemulsions are mixed with cellulose pulp, and applied as poultices over a thin sheet of Japanese paper (the paper grammage may range from 8.6 g/m^2 to 11 g/m^2), whose role is to avoid the adhesion of the poultice to the partially solubilized coating. The microemulsion is mixed with the sorbent until a wet pulp, with good workable quality (spreadability) is achieved (e.g. the poultice loaded with the microemulsion is wet enough but not dripping). The wet pulp is manually distributed over the Japanese paper layer, as evenly as possible. The thickness of the poultice may range from few millimeters to 10–20 mm, depending on the amount of polymer coating to be removed.

Once the poultice has been placed, the application of a layer of plastic (e.g. Egapack®, Melinex®, or others) or aluminum wrap over the poultice is recommended when the evaporation rate of the liquid components needs to be controlled. For instance, this is the case when relative humidity is significantly low (RH<40%). Obviously, the covering film (e.g. Melinex® or others) must not be soluble in the cleaning fluid (microemulsions, micellar solutions).

The application time may range approximately from 60 to 180 min. Typically, 1 h is necessary to have a complete action. The poultice must remain wet because letting the microemulsion poultice dry completely might make the removal of the polymer (and poultice) more difficult, as the wet swollen/softened coating might start to dry again. Two shorter applications are usually preferred to a single application longer than 2 h.

The mechanism of polymers removal depends strongly on their chemical nature. Some classes of organic materials are fairly soluble into the droplets of microemulsions or perfectly emulsifiable (e.g. wax and dammar); the solubilized material migrates into the poultice. In several other cases, the film swelling and the de-wetting of the surface (i.e. a process that modifies its wettability) drive the detachment of the coating. Therefore, the application time of nanostructured fluids must be carefully determined after experimental tests, in order to maximize the swelling/softening of the adhesives. Eventually, the poultice is removed and the remaining material (swollen polymer) is removed with a gentle mechanical action using a cotton swab embedded with distilled water.

Immediately upon removal of the poultice, rinsing with water is recommended, in order to eliminate possible residues of surfactant (the only non-volatile component of the cleaning system). However, depending on the technique originally used by the artist (fresco, "secco", tempera, etc.) the cleaning with water should be carried out carefully and different methods can be used. In case of fresco paintings in good conservation conditions, a sponge wet with water can be used to rinse the fresco until the possible remaining surfactant is completely removed. The mechanical action of the sponge might cause profuse lathering, even though the amount of residues is very low (few milligrams). In case of flaking paintings, the mechanical action of the sponge might be harmful and the flaking pigments could be substantially removed from the painting. In order to avoid this the Japanese paper should not be removed from the wall and the cleaning should be performed with a wet Arbocel® compress applied on the Japanese paper, and removed only when dry. During the drying process the usually very low amounts of surfactant remaining on the surface migrate into the wet cellulose pulp. This process can be repeated until no surfactant is detected in the cellulose pulp by an appropriate analytical technique. Usually 2–3 rinsing cycles are enough to remove all the surfactant. When the pigments are flaking it is mandatory to consolidate the painted layer using for instance nanoparticles as highlighted in the previous paragraphs and detailed in Chap. 2. The use of biodegradable non-ionic surfactants (mentioned above) minimizes the necessity of rinsing with water. The absence of residues of surfactants on treated paintings and stone (following the application of microemulsions/micellar solutions and rinsing with water) can be checked by analyses such as chromatography and FTIR (Grassi et al. 2007, 2009).

The main practical steps for the application of microemulsions and micellar solutions with the poultice technique have been resumed in Fig. 3.8a,b.

Different sorbent materials widely used in wall paintings conservation have been tested for loading with nanostructured fluids. Among them, cellulose powder of two different fiber lengths are the most suitable, i.e. Arbocel® BC200 (average size 0.3 mm) and Arbocel® BWW40 (average size 0.2 mm). In particular, Arbocel® BC200 has also been broadly used for the removal of salts (the extraction of soluble sulfates from wall paintings using the Ferroni method has been detailed in Chap. 2).

Fig. 3.8a The main practical steps for the application of cellulose poultices loaded with microemulsions (or micellar solutions) for the removal of detrimental polymer coatings from the surface of wall paintings (see text for further details). Part I. *1,2* The microemulsions are mixed with cellulose pulp, and applied as poultices over a thin sheet of Japanese paper. *3* The application time may range approximately from 60 to 180 minutes. *4* Eventually, the poultice is removed and the remaining material (swollen polymer), if any, is removed with a gentle mechanical action *5* using a cotton swab embedded with distilled water (if the conservation condition of the painting allows a gentle mechanical action; if not, partial consolidation is necessary before re-applying the microemulsion to the painting. Consolidation and cleaning cycles can be repeated several times, see Chapter text for details). *6* The cleaned area. (see Fig. 3.8b for the following steps). (Image by Michele Baglioni)

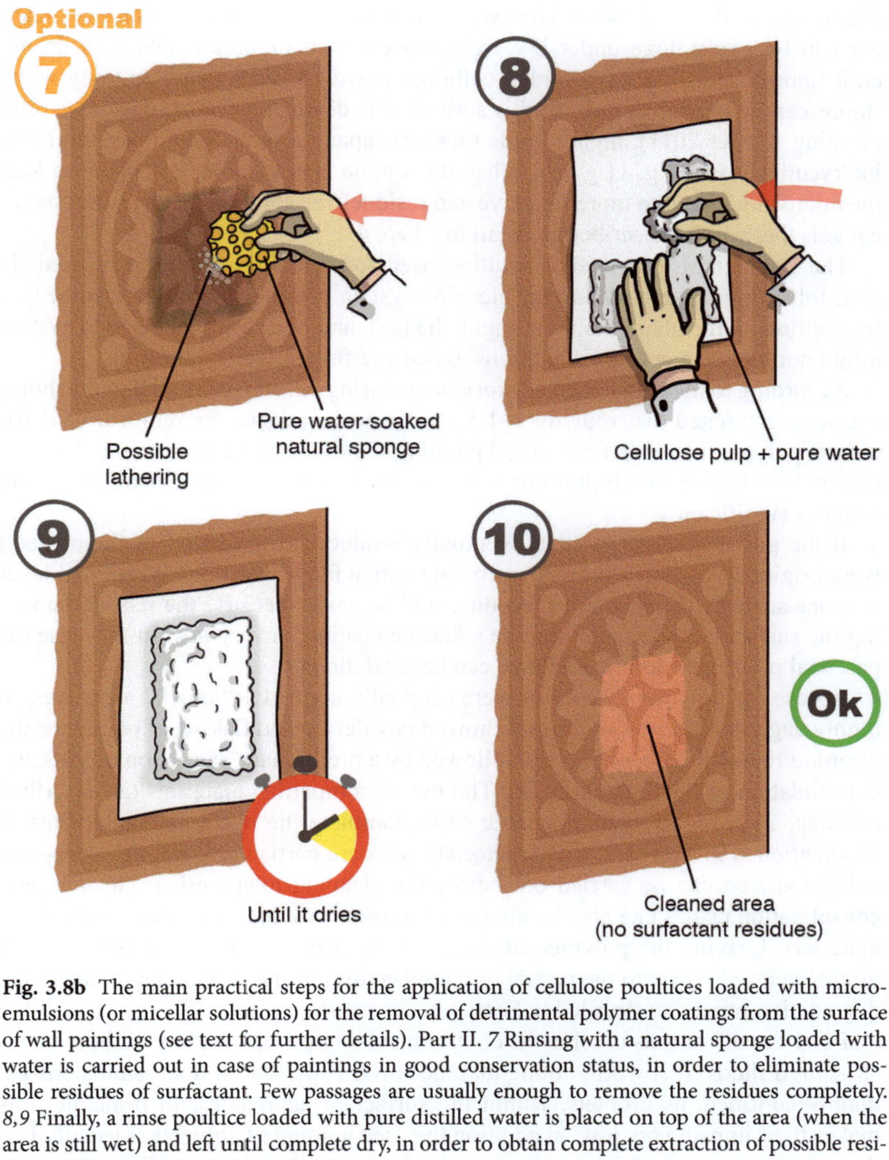

Fig. 3.8b The main practical steps for the application of cellulose poultices loaded with microemulsions (or micellar solutions) for the removal of detrimental polymer coatings from the surface of wall paintings (see text for further details). Part II. *7* Rinsing with a natural sponge loaded with water is carried out in case of paintings in good conservation status, in order to eliminate possible residues of surfactant. Few passages are usually enough to remove the residues completely. *8,9* Finally, a rinse poultice loaded with pure distilled water is placed on top of the area (when the area is still wet) and left until complete dry, in order to obtain complete extraction of possible residues *10*. (Image by Michele Baglioni)

The poultice technique is difficult to apply on concave surfaces (e.g. vaults) because poultices tend to detach after application (Brajer 2011). A possible proposed solution is to add materials with shorter fiber length so to enhance adhesion.

Residues of cellulose pulp might remain on the treated surface; however, no damage to paintings ascribable to pulp residues have been observed following the application of microemulsions or micellar solutions on wall paintings with the poultice technique detailed above. The examination and evaluation of the surface cleaning can be easily done under UV light, since carbonate-based substrates do not emit fluorescence. On the contrary, cellulose powder residues emit a strong white fluorescence. In case the use of this sorbent was deemed an issue, despite careful cleaning (Brajer 2011), higher grade (denser) Japanese tissue could be used as an intervention layer (e.g. 11 g/m^2). When this option is not viable, it is better to load the microemulsion into more retentive and residue-free sorbents, such as the chemical gels that will be described in detail in Chap. 4.

The microemulsion/micellar solution used for cleaning cannot be re-used: in fact, following the application the cleaning system composition might change (i.e. its position on the phase diagram might change), and as a result the re-used system might not be in the optimal conditions to achieve fully effective cleaning.

According to the data collected worldwide during several restoration workshops, it can be estimated that roughly 1–1.5 l/m^2 is necessary for the removal of detrimental polymer coatings from mural paintings. Built heritage materials and monuments show higher variability; this is because the porosity of the substrates usually changes significantly.

If the polymer coatings (that eventually produced degradation of the artifact) were originally used as a protective coating onto a fairly well cohered painted layer or stone surface, the cleaning procedure will be easier because the risk of damaging the surfaces, while removing the adhesive coating, is very low. In this case the practical procedure described above can be used directly.

In case the polymeric coatings were originally applied as "quick" adhesives on highly degraded painted layers that showed powdering and flaking of pigments, the cleaning/removal process must be followed by a preliminary extraction of salts and consolidation of the painted layers. The use of compatible materials (as described in Chap. 2) is highly recommended, so to complete the restoration intervention. As mentioned in Sect. 3.1, when pictorial layers are particularly fragile, a pre-consolidation step can be carried out before the cleaning intervention; cleaning and consolidation cycles can also be alternated to minimize the loss of original artistic materials. Leaving the polymer adhesive on the surface is not advisable, since it might eventually lead to the partial or complete loss of the painted layer due to the degradation processes detailed in Sect. 3.1.

Final considerations involve the environmental impact and safety of these nanostructured fluids. It is worth noting that the solvents present in microemulsions are mainly confined, as droplets, within the surfactant aggregates. In principle, this makes their impact very low. In addition the used amount of solvents is usually less than 10–15 % (w/w) of the whole cleaning fluid, and this also contributes to minimize both environmental and health impact. On the other hand, the non-volatile part is constituted by surfactants and these compounds (if non-biodegradable) may present some concern when dispersed into the environment. In spite of that, the application of the poultice technique (and rinsing steps as explained above) allows a complete removal of surfactants through extraction; therefore, the discarding of the used contaminated poultice represents the only relevant issue. The use of bio-degradable surfactants, as previously mentioned, contributes to minimize this concern.

The handling of microemulsions requires, as a good laboratory practice, the use of gloves and goggles. The smell of these systems (if any), due to the organic volatile components, may be fastidious, even when the organic content is extremely low. As a general rule, the use of microemulsions requires some ventilation, even though the toxicity impact tends to zero.

References

Azemar N (1997) The role of microemulsions in detergency processes. In: Solans C, Kunied H (eds) Industrial applications of microemulsions. Marcel Dekker Inc, New York, p 375

Baglioni M, Rengstl D, Berti D, Bonini M, Giorgi R, Baglioni P (2010) Removal of acrylic coatings from works of art by means of nanofluids: understanding the mechanism at the nanoscale. Nanoscale 2:1723–1732

Baglioni M, Giorgi R, Berti D, Baglioni P (2012a) Smart cleaning of cultural heritage: a new challenge for soft nanoscience. Nanoscale 4:42–53

Baglioni M, Berti D, Teixeira J, Giorgi R, Baglioni P (2012b) Nanostructured surfactant-based systems for the removal of polymers from wall paintings: a small-angle neutron scattering study. Langmuir 28:15193–15202

Baglioni P, Chelazzi D, Giorgi R, Poggi G (2013) Colloid and materials science for the conservation of cultural heritage: cleaning, consolidation, and deacidification. Langmuir 29:5110–5122

Baglioni P, Berti D, Bonini M, Carretti E, Dei L, Fratini F, Giorgi R (2014) Micelle, microemulsions, and gels for the conservation of cultural heritage. Adv Colloid Interface Sci 205:361–371

Borgioli L, Caminati G, Gabrielli G (1995) Removal of hydrophobic impurities from pictorial surfaces by means of heterogeneous systems. Sci Technol Cult Herit 4:67–74

Brajer I (2011) Testing non-standard cleaning methods on the wall paintings in Undløse Church. Unpublished report, Nr. 11073410. Archive of the Department of Conservation, National Museum of Denmark

Burke J (1984) Solubility parameters: theory and application. In: The book and paper group annual, vol III. The American Institute for Conservation. http://cool.conservation-us.org/coolaic/sg/bpg/annual/v03/bp03–04.html. Accessed 4 June 2014

Carretti E, Giorgi R (2013) Cleaning IV: applications and case studies. In: Baglioni P, Chelazzi D (eds) Nanoscience for the conservation of works of art. RSC, London, pp 280–314

Carretti E, Dei L, Miliani C et al (2001) Oil-in-water microemulsions to solubilize acrylic copolymers: application in cultural heritage conservation. Progr Colloid Polym Sci 118:63–67

Carretti E, Dei L, Baglioni P (2003) Solubilization of acrylic and vinyl polymers in nanocontainer solutions. Application of microemulsions and micelles to cultural heritage conservation. Langmuir 19:7867–7872

Carretti E, Giorgi R, Berti D, Baglioni P (2007) Oil-in-Water nanocontainers as low environmental impact cleaning tools for works of art: two case studies. Langmuir 23:6396–6403

Cedillo L (1991) La conservación en zonas arqueológicas. Tres décadas de trabajo. Bachelor thesis in Conservation, ENCRyM-INAH, Mexico

Chevalier A, Chelazzi D, Baglioni P et al (2008) In: Proceedings of the 15th Triennal conference ICOM committee for conservation, vol I, New Delhi, pp 581–589

Chiantore O, Lazzari M (2001) Photo-oxidative stability of paraloid acrylic protective polymers. Polymer 42:17–27

Danielsson I, Lindman B (1981) The definition of microemulsion. Colloids Surf 3:391–392

Estoe J (2005) Microemulsions. In: Cosgrove T (ed) Colloid science. Blackwell Publishing Ltd., Oxford, UK.

Evans DF, Wennerström H (1994) The colloidal domain: where physics, chemistry, biology, and technology meet. VCH, New York

Evans DF, Wennerström H (1999) The colloidal domain: where physics, chemistry, biology, and technology meet (2nd edn). Wiley, New York

Favaro M, Mendichi R, Ossola F et al (2006) Evaluation of polymers for conservation treatments of outdoor exposed stone monuments. Part I: photo-oxidative weathering. Polym Degrad Stab 91:3083–3096

Ferroni E, Gabrielli G, Caminati G (1992) ASportazione di materiali idrofobi da superfici pittoriche murali mediante solubilizzazione in sistemi dispersi. In: Olivetti (ed) La Cappella Brancacci, la scienza per Masaccio, Masolino e Filippino Lippi. QUaderni del restauro, Milan, p 162–171

Ferroni E, Gabrielli G, Caminati G (1992) In: La Cappella Brancacci, la scienza per Masaccio, Masolino e Filippino Lippi. Quaderni del restauro, Milan (Olivetti Eds)

Giorgi R, Baglioni M, Berti D et al (2010) New methodologies for the conservation of cultural heritage: micellar solutions, microemulsions, and hydroxide nanoparticles. Acc Chem Res 43:695–704

Goodwin J (2009) Colloids and interfaces with surfactants and polymers, 2nd edn. Wiley, Chichester

Grassi S, Carretti E, Pecorelli P et al (2007) The conservation of the Vecchietta's wall paintings in the old sacristy of Santa Maria della Scala in Siena: the use of nanotechnological cleaning agents. J Cult Herit 8:119–125

Grassi S, Favaro M, Tomasin P et al (2009) Nanocontainer aqueous systems for removing polymeric materials from marble surfaces: a new and promising tool in cultural heritage conservation. J Cult Herit 10:347–355

Holmberg K, Jonsson B, Kronberg B et al (2002) Surfactants and polymers in aqueous solutions, 2nd edn. Wiley, Chichester

Horie CV (1987) Materials for conservation: organic consolidants, adhesives, and coatings. Butterworths, London

Lange KR (1999) Surfactants: a practical handbook. Hanser Gardner Publications, Cincinnati

Lawrence MJ, Rees GD (2000) Microemulsion-based media ad novel drug delivery systems. Adv Drug Deliv Rev 45:89–121

Lazzari M, Chiantore O (2000) Thermal-ageing of paraloid acrylic protective polymers. Polymer 41:6447–6455

López Ortega T (1999) Evaluación de los tratamientos de conservación efectuados en los mascarones de estuco de Kohulnich, Quintana Roo. Bachelor thesis in conservation, ENCRyM-INAH, Mexico

Luders H (2000) Synthesis of alkyl glucosides and alkyl polyglucosides in nonionic surfactants. In: Balzer D, Luders H (eds) Alkyl polyglucosides. Marcel Dekker, New York, pp 7–76

Magaloni Kerpel D (2011) La pintura mural y su conservación. Arqueología Mexicana 108:33–37

Mora P, Mora L, Philippot P (1984) Conservation of wall paintings. Butterworths, London

Palazzo G, Fiorentino D, Colafemmina G et al (2005) Nanostructured fluids based on propylene carbonate/water mixtures. Langmuir 21:6717–6725

Philippot P, Mora P (1969) La conservación de pinturas murales, in La conservación de los bienes culturales con especial referencia a las condiciones tropicales. UNESCO, Paris, pp 181–202

Pizzorusso G, Fratini E, Eiblmeier J et al (2012) Physicochemical characterization of acrylamide/bisacrylamide hydrogels and their application for the conservation of easel paintings. Langmuir 28:3952–3961

Rance G (1977) Micellar solutions versus microemulsions. J Colloid Interface Sci 60:207–209

Stubenrauch C (2009) Microemulsions: background, new concepts, applications, perspectives. Wiley-Blackwell, Chichester

Chapter 4
Cleaning of Easel Paintings

Abstract Cleaning of easel paintings is traditionally carried out using organic solvents or aqueous solutions. One of the main limitations is due to the lack of control in the cleaning action when non-confined solvents and solutions are used. Undesired effects include swelling or solubilization of original artistic materials, uncontrolled or excessive removal of layers and patinas, and re-transport of solubilized matter through the porous matrix of the artifact. These issues can be effectively addressed by confinement of the cleaning fluids in gel networks. As a matter of fact, conservators have adopted several gel and 'gel-like' formulations (e.g. viscous dispersions of thickening agents), aiming at the non-invasive and gradual cleaning of works of art. This chapter reports on the main traditional systems and on the innovative gels and dispersions developed in the last decade in the framework of colloid and surface science for the preservation of cultural heritage. In particular, both highly retentive chemical gels and "peelable" viscoelastic dispersions are discussed, illustrating applicative procedures.

4.1 Background

Cleaning of paintings is a delicate operation, since it poses risks of altering the original artistic substrates. Typically, the term "cleaning" refers both to "surface cleaning" and to the removal of degraded layers (Cremonesi and Signorini 2012). In the first case grime and soil are removed, ideally without affecting any film-forming material (varnishes, retouching, painted layers). In the second case, the intervention aims at the partial or complete removal of altered and degraded layers, such as yellowed varnishes, aged and/or detrimental coatings, etc.

Through the ages, restorers and conservators have used a vast number of cleaning materials, including soaps, alimentary products, inorganic materials and even biofluids such as saliva. Presently, the cleaning of artifacts is still largely carried out using organic solvents, also coupled with mechanical action. However, the application of solvents can lead to undesired effects such as the swelling of binding media, the alteration of pigments, the transport of dissolved matter through porous

P. Baglioni et al., *Nanotechnologies in the Conservation of Cultural Heritage*,
DOI 10.1007/978-94-017-9303-2_4

matrices, or simply the lack of control in the cleaning process, resulting in the removal of original layers or "patinas" that are sometimes considered as historical parts of the artifacts. Moreover, the use of solvents involves specific risks for the operator's health that must be considered.

The use of aqueous solutions (e.g. of chelating agents, enzymes, surfactants, acids, bases) has gained increasing attention in the last decades due to health considerations and to the possibility of controlling in a relatively safe way both the cleaning effectiveness and the interaction with the paint binding media, playing on factors such as the solutions viscosity, pH and ionic strength (Wolbers 2000; Casoli et al. 2014). Moreover, water-based nanostructured cleaning fluids such as oil-in-water microemulsions and micellar solutions have proved to be valid alternatives to solvents for the removal of undesired materials, detrimental coatings and varnishes (Baglioni et al. 2014), as detailed in Chap. 3.

However, the direct use of aqueous solutions and water-based systems is discouraged on water-sensitive substrates (canvas, preparation layers, paper artworks) since it might lead to undesired effects such as the swelling of fibers or hydrophilic layers, and the leaching or solubilization of soluble components.

All the aforementioned issues can be efficiently addressed by confining the cleaning fluids within a gel matrix (polymer network) that releases them in a controlled way onto the surface of the work of art. Fluids can be thickened ("gelled") or directly loaded into gels, which introduces several advantages:

- The slow release of solvents, solutions or emulsions across the gel-artifact interface leads to controlled cleaning at a range of different rates.
- The uncontrolled diffusion of the cleaning fluids and of dissolved grime through the layers of the artistic substrate is partially or completely avoided.
- The evaporation rate of confined fluids is drastically decreased; therefore the toxicity of the cleaning systems is reduced.

In fact, research work in the field of cultural heritage conservation has extensively explored the use of gels as highly effective and versatile tools for a wide number of applications and the role of these materials in the cleaning of works of art has become central. In this framework, colloid and surface science have greatly contributed to the development of innovative systems with specific features that maximize the control of the cleaning action. In the following paragraphs we provide an overview of the main gel systems that have been used for the cleaning of easel paintings, and of the systems developed at CSGI (Consorzio Interuniversitario per lo Sviluppo dei Sistemi a Grande Interfase, Center for Colloid and Surface Science) focusing on advanced materials with enhanced properties.

There is a vast amount of literature on gels, and a detailed description of these systems is beyond the scopes of this book. In this chapter we will focus specifically on formulations used for the cleaning of artifacts, dealing with those aspects that are relevant for practical applications. In fact, the precise definition of "gel" is not a straightforward task, as it involves concepts derived from rheology and polymer science. A gel can be defined as a material that comprises a three-dimensional solid network immersed in a fluid, where the network is "percolative", i.e. only a single

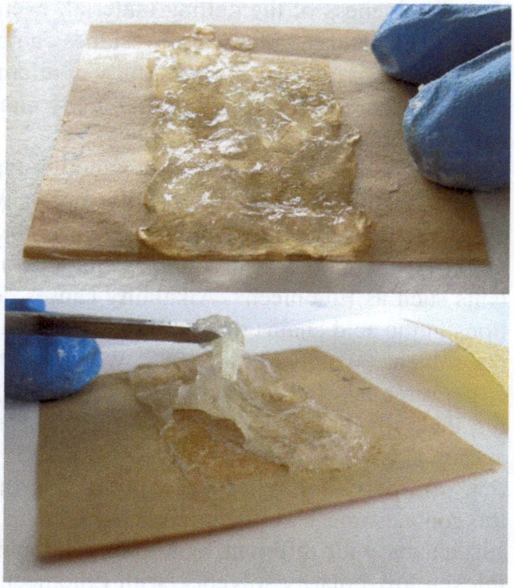

Fig. 4.1 A cellulose ether "gel" (Klucel®, 6%), loaded with alcohol, acetone and a hydrocarbon solvent. (Printed with the kind permission of SELIDO (at IPCE, Instituto del Patrimonio Cultural de España, Madrid))

aggregate can be found. In a larger sense, a gel could be defined tentatively as a soft material that consists of interconnected polymer chains entrapping a fluid (usually a liquid). The polymer is called "gelator" or "gellant". The term "hydrogel" is used when the liquid is water, while "organogel" indicates gels that confine organic solvents. Another important classification considers the type of interactions that build the network between the polymer chains. In "physical gels" the interactions are secondary bonds, such as dispersion forces, hydrogen bonds, electrostatic or hydrophobic interactions. Instead, in "chemical gels" the polymer network is formed by covalent bonds (e.g. cross-links between polymer chains), which are more than ten times stronger than a hydrogen bond and more than 1000 times stronger than dispersion forces.

One of the main classes of materials traditionally used as thickening agents for the cleaning of artifacts comprises cellulose ethers (e.g. Klucel®, Tylose, HPMC), which are widely known in cosmetics, pharmaceutical and food industry. Although these systems cannot be strictly considered as gels based on their rheological behavior, they macroscopically look as gels and are commonly referred to as such for practical reasons. The preparation of "gels" using cellulose ethers is relatively easy: the polymers are dispersed in water, stirring either at room temperature or using heat, and letting rest for about one day. These polymers can be used to thicken water and, in the case of Klucel G®, mainly polar solvents such as some alcohols. The viscosity of cellulose ether "gels" depends on the polymer DP and on the solvent loaded. Usually, pasty gel-like dispersions are obtained, which exhibit film forming and adhesive properties (see Fig. 4.1). Cellulose ether "gels" are sensitive to pH values either too acidic ($pH<3$) or too alkaline ($pH>11$). Moreover, these physical gel-like networks are prone to leaving residues that are hard to remove, due to the adhesiveness of the gelling material, even when rinsing is carried out (Casoli et al. 2014), and the removal of residues can be invasive for the artifact. Cellulose residues might increase the possibility of microbial proliferation on the artifact, also depending on the substrate hydrophilicity and on the degree of substitution of the cellulose polymer.

The so-called "solvent gels" are another type of materials traditionally used for the cleaning of easel paintings. Solvent gels were developed in the late 1980s and early 1990s by Richard Wolbers, and are obtained using thickeners derived from polyacrylic acid (e.g. Carbopol®, Pemulens®) and weakly basic non-ionic surfactants such as Ethomeen®. Alkalinity causes deprotonation of the carboxylic functions in the acid chains that unfold forming an extended 3D network and confining the solvent (Wolbers et al. 1988; Wolbers 2000). The surfactant properties allow easier removal of grime and undesired layers. Solvent gels are prepared by dispersing the gellant (polyacrylic acid) into a solvent, and adding the surfactant as the system is under stirring. Finally, water (ca. 5 % w/w) is added to induce the formation of a pasty gel. After a few hours rest, the gels can be applied using a swab, the contact time with the paint surface being of 1–5 min. By selecting Ethomeen surfactants with different hydrophilic–lipophilic balance (HLB), it is possible to thicken low-polarity solvents (with Ethomeen C12®) or polar solvents (with Ethomeen C25®). Solvent gels can also be used to control the action of detergents and enzymes. In fact, these systems are still among the most used cleaning tools due to their effectiveness and versatility.

However, as for other physical gels and gel-like dispersions, one of the main drawbacks is related to the removal of residues of the cleaning system (gellant and non-volatile components) after the application. Burnstock and Kieslish (1996) demonstrated, through gas chromatography/mass spectroscopy (GC-MS) and scanning electron microscopy (SEM), that residues of Ethomeen are present on the surface of canvas paintings both after mechanical removal of the solvent gels using a swab roll, and after applying clearing solvents. Moreover, SEM examination suggested that the solvents used for removing the residues influenced the surface appearance of the painting, and similar surface changes were observed using solvents alone for varnish removal. Later, Burnstock and White (2000) assessed the nature of the degradation products of Ethomeen C-12 residues (left by solvent-gel treatment) after both short-term and long-term light aging. The oxidation products of the surfactant are formed over medium- to long-term aging. The authors also considered an "extreme" scenario by aging mixtures of di- and triterpenoid natural resins with Ethomeen, and found that the presence of the surfactant has a synergetic influence on the degradation processes. The nature of Ethomeen degradation products (including amine N-oxides) raised some concern about long-term (e.g. several decades) contact of the residues with oil/resin paint media. Stulik et al. (2004) carried out an extensive study on the theory and application of solvent gels, including the detection of residues on the surfaces of the treated artifacts, the stability of surfactants during natural and accelerated aging, and recommendations for formulating gels for specific tasks.

Fig. 4.2 An agar gel (2 %) loaded with water. (Printed with the kind permission of SELIDO (at IPCE, Instituto del Patrimonio Cultural de España, Madrid))

Some polysaccharide materials such as agar, gellan and xanthan gum have been proposed in the last years for the surface cleaning of artifacts, to be used either as highly viscose solutions or as "rigid" gels that retain the shape of the container where they are prepared. Agar and gellan are used as gellants in biomedicine, pharmacology and food industry. These materials are non-toxic and can be attacked by some bacteria and fungi (agar is indeed used in biology for the preparation of culture media). Xanthan gum is also used in food industry as a rheology modifier and stabilizer.

Agar is extracted from red seaweeds (mainly *Gelidium* and *Gracilaria* genera), and is composed of agaropectin and agarose. Agar is used as a thickening agent, as electrode binder for electrolyte cells, and for drug delivery. When mixed with water (concentrations range from 1 to 4 %) and heated up to 80 °C it forms random coils, which then rearrange upon cooling (below 40 °C) forming a thermo-reversible highly porous 3D gel structure where double helix chains are linked through hydrogen bonding (Davies et al. 2010). Agar gels can be used with chelating agents, enzymes or surfactants, at different pH values, and can be applied either as a poultice (e.g. by brushing) or as a rigid gel (see Fig. 4.2), leaving almost no residues as reported in the literature (Gulotta et al. 2014; Cremonesi 2006). Recently, agar gels have also been loaded with nanostructured cleaning fluids (microemulsions), for the removal of lipophilic materials (wax) from porous surfaces (Gorel 2010).

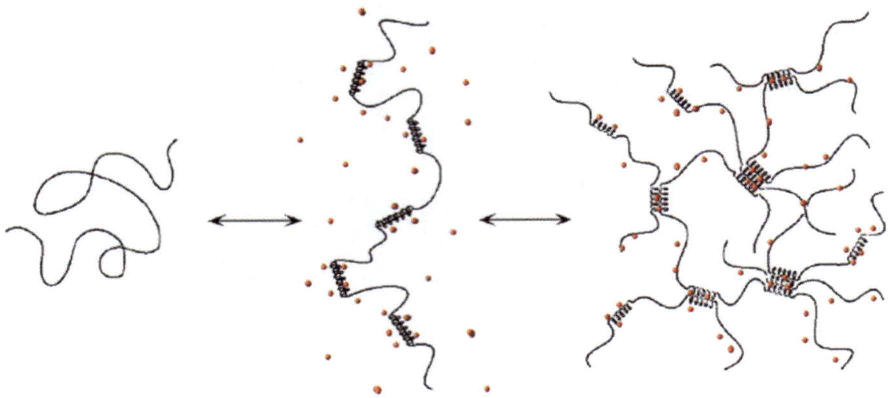

Fig. 4.3 Scheme of the coil-helix transition in gellan. Ca^{2+} ions stabilize the gel structure. (Repreinted with permission from Caggioni et al. (2007))

Gellan is a linear anionic polymer produced by the bacterium *Sphingomonas elodea*. The repeating unit of the polymer is a tetrasaccharide consisting of (1–3)-β-D-glucose, (1–4)-β-D-glucuronic acid, (1–4)-β-D-glucose and (1–4)-α-L-rhamnose. Figure 4.3 shows a scheme of the gel formation process. The mechanism is similar to that described above: upon heating, the polymer is present in aqueous solutions as coils. Cooling down to 30–40 °C, the molecules form double helices that link through secondary bonds (hydrogen bonds, Van der Waals interactions), randomly alternated with coils. Cations, in particular Ca^{2+} ions (for instance added as calcium acetate solutions) stabilize the structure (see Fig. 4.3). Both elastic (see Fig. 4.4) and brittle gels can be formed depending on cation concentrations and on the acyl content of the polymer (Mao et al. 2001; Bajaj et al. 2007). For the cleaning of artifacts, it is possible to obtain a compact, visco-elastic, non-adhesive and homogeneous structure that is stable at different pH values. Gellan gels are more transparent than agar, and exhibit higher water retention at concentrations of 1–2 %. Iannucelli and Sotgiu (2010) used gellan for the cleaning of paper artworks, where it was observed that the gels gradually release water onto the paper substrate and pick up dirt (the gels yellow following application), and can be removed without leaving substantial residues as opposed to cellulose ether gels. The authors also showed that it is possible to prepare gels loaded either with an aqueous enzymatic solution or with *tert*-butyl aminoborane, which has been used as a mild reagent to reduce carbonyl groups in oxidized paper (Bicchieri and Brusa 1997; Bicchieri et al. 2000). In the first case, the enzymatic solution was loaded into the gellan after the gelification process, by dripping the solution directly over the gel surface with a micropipette and using a brush for distributing the solution. Instead, *tert*-butyl aminoborane was added after the polymer dispersion ended the heating cycle. Gellan gels can also be used with small amounts of solvents such as ethanol or isopropanol. Commercial products include Gelzan® and Kelcogel®.

Finally, Xanthan gum (e.g. Vanzan®) is extracted after fermentation of bacteria (*Xanthomonas campestris*). The polymer forms viscous gels that can be used in a

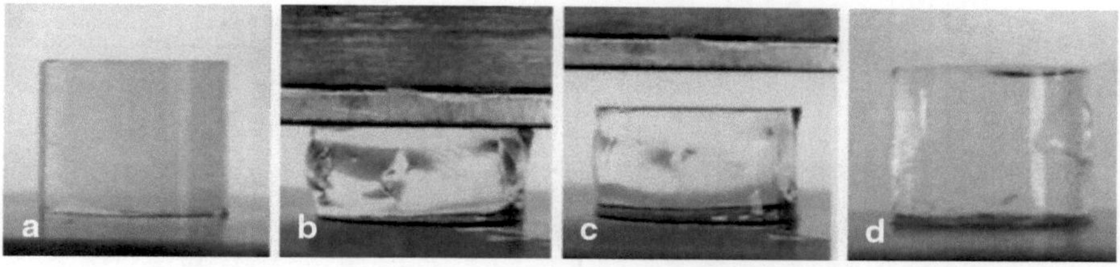

Fig. 4.4 Photographs of a gellan gel (1.33 %) compressed at 0.1 mm/min. **a** Undeformed gel. **b** A compressed gel by the cross head. **c** A compressed gel after removing the cross head. **d** The gel of (**c**) soaked in water for 2 days. (Reprinted from Nakamura K et al. (2001))

wide pH range, and are compatible with chelates, several solvents (gelling roughly 40 % aqueous solutions of alcohols or glycols) and emulsions. Also in this case, the gelling mechanism involves the formation of helices starting from random coils, and the obtained gel is thixotropic.

Gels such as those formed by agar and gellan can retain high amounts of water, however is some cases a higher water retention is required for the treatment of water-sensitive works of art, where leaching or loss of components (colors, inks) might take place due to excessive wetting (Domingues et al. 2013).

As a matter of fact, experimentation of new formulations with enhanced properties, as compared to the materials previously discussed, is continuously ongoing. Research work is centered around developing gels with key features such as improved mechanical properties (so to allow easy handling and completely avoid gel residues on the treated surfaces) and high retentiveness to fully control the cleaning action even on sensitive substrates. In the last decade, research work at CSGI has targeted these requirements, resulting in the development of several formulations that have been successfully used for the cleaning of artistic surfaces, including easel paintings. An example is that of responsive gels, i.e. a class of materials that respond to an external stimulus such as temperature, pH or magnetic fields for practical purposes (Bonini et al. 2007, 2008; Carretti et al. 2010a). Moreover, gels with high visco-elasticity that can be easily "peeled" off the surface, and highly retentive chemical gels for the cleaning of water-sensitive surfaces, have been realized and assessed on movable works of art (Baglioni et al. 2014). The following sections will provide a description of the main innovative formulations, including their application field and practical procedures for their use on easel paintings.

4.2 Formulations

Research work in the field of colloids and soft matter science has produced in the last decades a palette of advanced gels and gel-like systems with different features, for the cleaning of works of art.

The first responsive systems that were developed in this framework were rheoreversible gels and nanomagnetic sponges. Carretti et al. (2003, 2004) showed that solutions of polyamine (e.g. polyallilamine, PAA) form a gel by adding CO_2 (through

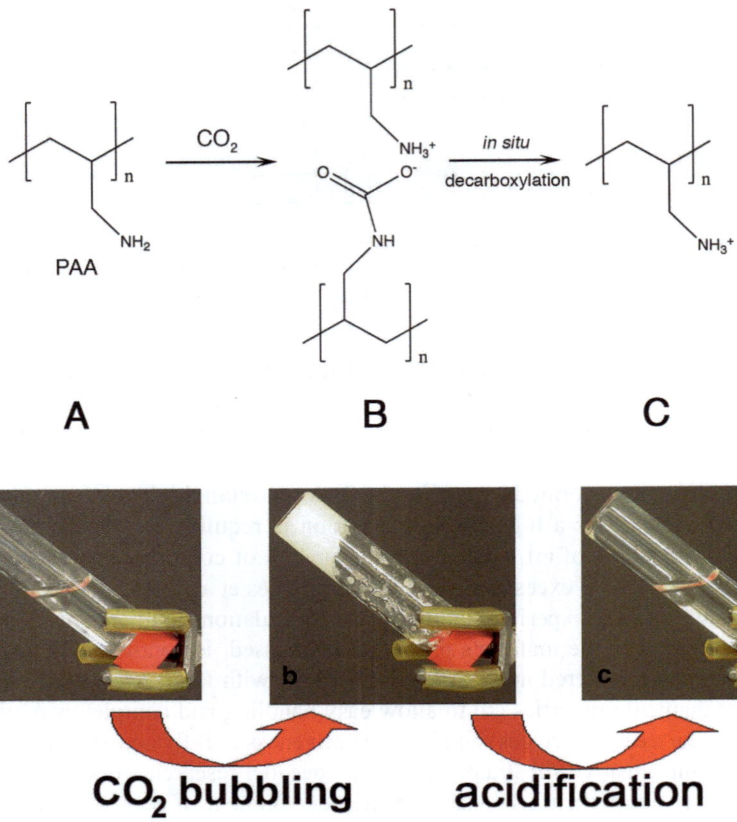

Fig. 4.5 Rheoreversibility of a polyallilamine-based gel. **a** A solution of polyallilamine (PAA) in 1-pentanol. **b** Bubbling CO_2 leads to the formation of ammonium and carbamate groups, and the PAA·CO_2 gel is formed due to electrostatic interactions. **c** The addition of a weak solution of acetic acid causes the gel to revert to a fluid state. (Readapted with permission from Carretti et al. (2004))

bubbling). The 3D network is due to the formation of ammonium and carbamate groups on the polymer chains, which are attracted through electrostatic interactions as shown in Fig. 4.5b. The formed PAA·CO_2 gel can then be applied on painted surfaces. After the cleaning action, the PAA·CO_2 is decarboxylated (Fig. 4.5c) simply by adding in situ a small amount of a weak (0.05 M) aqueous solution of acetic acid. As a result, the viscoelastic gel reverts back to a liquid, which is wiped away with a cotton swab. For instance, the gel was applied to remove an undesired lacquer layer, applied during past restoration interventions, from the surface of a fourteenth century egg tempera wood painting (from the National Gallery of Siena, Italy), see Fig. 4.6. The gel was obtained starting from a solution of PAA in 1-pentanol, and was left in direct contact with the paint for 1 min. Then, a few microliters of weak acetic acid solution was added and the liquefied mixture was removed with a swab. As confirmed by analysis through SEM-EDX, the use of gel allowed the removal of the lacquer. These systems were further assessed, for instance for the removal of

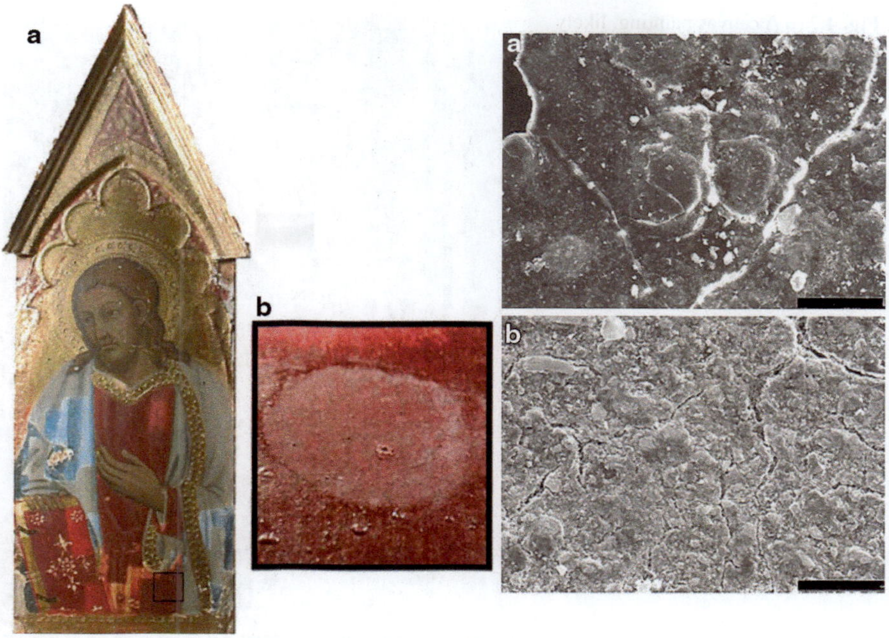

Fig. 4.6 **a** (*left*) A fourteenth century egg tempera wood painting (from the National Gallery of Siena, Italy). The black box indicates the area selected for the removal of an undesired lacquer layer, applied during past restoration interventions. **a** (*right*) A SEM picture of the surface before the removal of the lacquer (bar is 100 μm). **b** (*left*) Grazing light image of the painting surface, where the lacquer was removed using a PAA·CO_2/1-pentanol gel. **b** (*right*) The SEM analysis of the surface after the application of the gel shows a surface texture that is similar to that of paintings unaltered by organic coatings (bar is 100 μm). (Readapted with permission from Carretti et al. (2004))

aged varnish from the surface of a gilded nineteenth century frame (Carretti 2005). Later, polyethylenimines (PEI) were also used to make rheoreversible gels for the cleaning of painted surfaces and gilded wood (Carretti 2005, 2008, 2010a). For instance, a PEI solution of 1-octanol/xylenes was gelled by bubbling CO_2, and the PEI·CO_2 gel was used to successfully remove a glossy overpaint and varnishes from the surface of a canvas painting (contact time < 10 min), see Fig. 4.7. Stereomicroscopic analysis confirmed that the undesired layers had been effectively removed, with no skinning or abrasion of the original surface. Alternatively, PEI was used to gel solvents such as 1-methyl-2-pyrrolidone (MP) or alcohols, for instance 1-pentanol was gelled to remove a surface layer of degraded varnish from a painted wood sculpture (Fig. 4.8) (Baglioni et al. 2009).

The use of PPA- and PEI-based gels allowed excellent cleaning results and provided an example of the potentiality of colloids and soft matter science for the development of innovative and effective tools. Successive research work focused

Fig. 4.7 **a** A canvas painting, likely dating from the late nineteenth century. The painting had been lined, varnished and overpainted during restoration interventions. **b** Surface overpaint and varnishes were removed, using a rheoreversible PEI·CO_2 gel. (Reprinted with permission from Carretti et al. (2010a))

on different classes of responsive gels to further control the handling and removal of the cleaning systems.

Bonini et al. (2007, 2008) synthesized and characterized magnetic nanoparticles that were functionalized and associated with acrylamide-based gels, creating a nanomagnetic sponge (Fig. 4.9). Acrylamide exhibits good water loading and retention properties, which make these chemical hydrogels useful for loading aqueous solutions and water-based systems (e.g. microemulsions) and releasing them in a controlled way onto artistic substrates. The magnetic properties of the gel allow its easy and quick removal simply by using a permanent magnet (Fig. 4.10), so to completely avoid any direct handling of the gel during the removal step. In fact

Fig. 4.8 Eighteenth century painted wood sculpture with a surface layer of a degraded natural varnish before (*right side*) and after (*left side*) cleaning by a gel of poly(ethylene imine) as the gellant and 1-pentanol as the continuous phase. (Reprinted with permission from Baglioni et al. (2009))

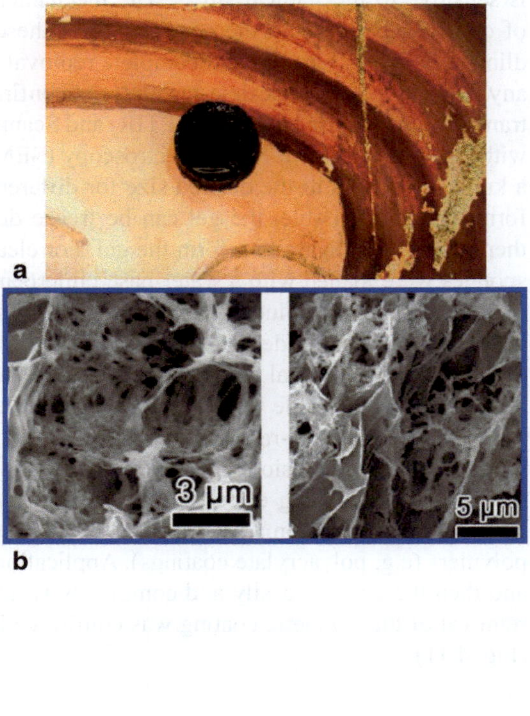

Fig. 4.9 **a** An acrylamide-based nanomagnetic gel on the surface of a painting. Reproduced with permission from Bonini et al. (2007) Nanomagnetic sponges for the cleaning of works of art. Langmuir 23:8681–8685. Copyright 2007 American Chemical Society. **b** SEM micrographs of the nanomagnetic gel at two different magnifications. Reproduced with permission from Bonini et al. (2008)

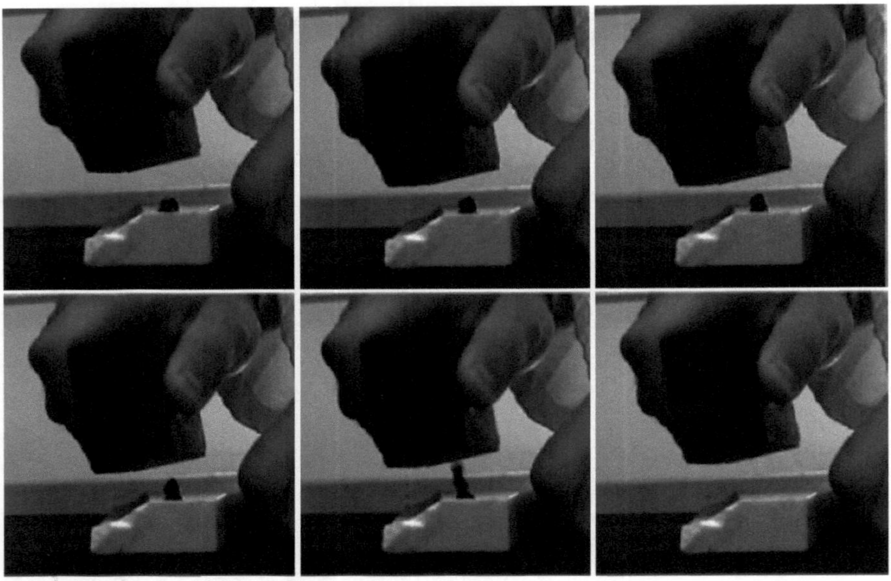

Fig. 4.10 Removal of the microemulsion-loaded nanomagnetic gel from a substrate using a permanent magnet. (Reproduced with permission from Bonini et al. (2007))

this tool is particularly suited to the cleaning of precious objects where the surface is sensitive to mechanical stress. The mechanical properties (due to the network of chemical bonds) and the consistency of these "gummy" gels allow easy handling (e.g. with tweezers) and complete removal after application, without leaving any detectable gel or particle residues, as confirmed by analyses through Fourier transform infrared spectroscopy (FTIR) and Scanning Electron Microscopy coupled with Energy-dispersive X-ray spectroscopy (SEM-EDS). The gels can be cut with a knife or scissors to the desired size for different applications. Moreover, the gel formation is reversible: the gel can be freeze dried, obtaining a powder that can then be hydrated again to reform the gel. For cleaning purposes, the nanomagnetic sponges were loaded with a water-based microemulsion containing small amounts of *p*-xylene and nitrodiluent (less than 2% each), similar to the systems described in Chap. 3. Polyacrylamide gels have macropores with a size ranging from hundreds of nanometers to several microns; therefore the nano-sized droplets contained in the microemulsion are able to diffuse through the gel network. Scattering techniques such as Small Angle X-ray Scattering (SAXS) indicated that, upon loading, both the gel and the microemulsion retain their structure, which is fundamental for achieving the maximum cleaning effectiveness. The microemulsion-loaded gel was applied both on marble and damaged paintings covered by undesired coatings of synthetic polymers (e.g. polyacrylate coatings). Application time ranged from 10 min to 2 h, and then the gel was easily and completely removed as described above. The full removal of the synthetic coating was confirmed by FTIR and SEM-EDX analyses (Fig. 4.11).

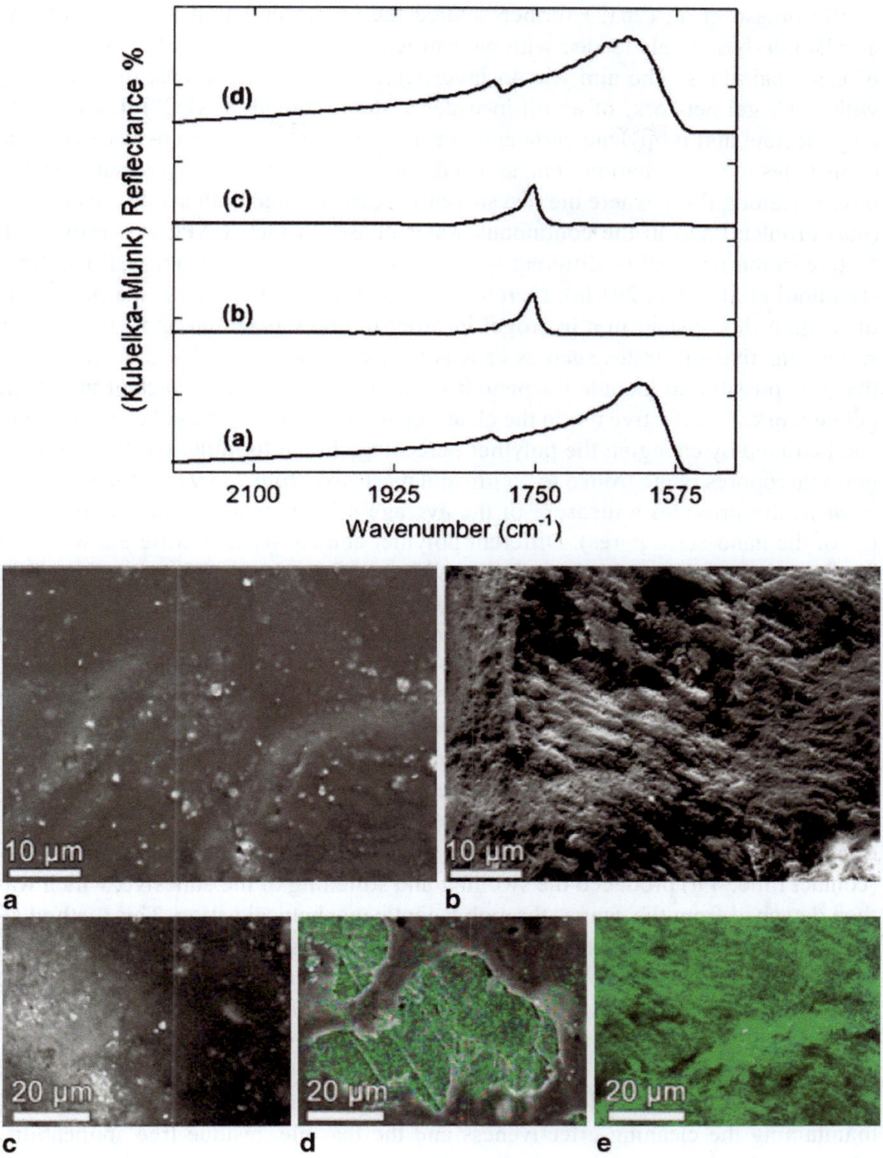

Fig. 4.11 (*Top panel*) Microreflectance FTIR spectra: (**a**) quarry marble sample without any treatment on the surface, (**b**) the same surface treated with a layer of Paraloid B72, (**c**) reference sample of Paraloid B72 as a pellet and (**d**) marble surface after complete removal of the Paraloid B72 layer using the nanomagnetic gel. The thick layer of Paraloid B72 completely hides calcium carbonate (calcite) bands (image **b**). The efficacy of the novel gel is highlighted by the complete disappearance of polymer and the recovery of the marble spectral features (image **d**). (*Bottom panel*) (**a**) SEMpicture of a marble surface treated with Paraloid B72 before cleaning and (**b**) after cleaning, (**c**) Mapping of calcium (green spot) through EDS (Cameo software) before cleaning, (**d**) 30 min after nanomagnetic gel application, and (**e**) after the complete removal of the polymer. Both panels are reproduced with permission from Bonini et al. (2007))

Pizzorusso et al. (2012) further studied the characterization and use of polyacrylamide-based gels loaded with nanostructured cleaning fluids for the cleaning of easel paintings. The aim was to investigate possible alterations, upon loading within the gel network, of an oil-in-water system (named “EAPC”) that contains ethyl acetate and propylene carbonate, beside a surfactant and a co-surfactant that contributes to the formation of nano-sized micelles. EAPC is a very versatile water-based cleaning fluid, where the two solvents are distributed both inside the micelles (nanodroplets) and in the continuous water phase. In fact, EAPC has proven effective in the removal of different types of detrimental coatings on wall paintings (Baglioni et al. 2010, 2012a); therefore, it was important to assess the possibility of loading this system in a hydrogel in order to allow potential applications onto water-sensitive substrates such as canvas paintings. SAXS analyses demonstrated that it is possible to include the nanodroplets of solvents and surfactant inside the gel network, for effective use in the cleaning of works of art. The gel characteristics can be tuned by changing the polymer percentage in the formulation. For instance, gels macropores range from 5 to 25 μm and mesh size from 7 to 9 nm (the mesh size intrinsically provides a measure of the average microporosity of the gel network, i.e. of the nano-sized pores). Different polymer content leads to different water retention properties, and the formulation with the highest retention was selected for the treatment of canvas in order to avoid uncontrolled release of the microemulsion across the surface (see Fig. 4.12). Acrylamide-based gels and microemulsions were applied on lined canvases to removed aged adhesives (Chevalier et al. 2008; Pizzorusso et al. 2012). For instance, gels loaded with EAPC were used on canvas samples that had been treated with Mowilith DM5®, a commercial adhesive based on vinyl acetate and *n*-butyl acrylate. In fact, polyvinyl acetate-based adhesives such as Mowilith DM5® and DMC2® are prone to degradation upon aging (Chelazzi et al. 2014), and one of the main degradation routes involves the release of organic acids that are detrimental to the canvas fibers. The application of the gel (contact time: 4 h) produced the swelling and softening of the adhesive, which was then detached from the canvas through a gentle mechanical action. This method allowed the perfectly localized action of the cleaning system, leading to the removal of most of the adhesive from the canvas without any uncontrolled spreading of the water-based EAPC fluid. Moreover, no detectable gel residues were present on the cleaned canvas surface, as assessed through FTIR analysis (Fig. 4.13).

Successively, research focused on chemical hydrogels with enhanced retentiveness, to allow the controlled cleaning of highly sensitive painted surfaces while maintaining the cleaning effectiveness and the feasible, residue-free applicability of the acrylamide-based formulations. Domingues et al. (2013, 2014) developed semi-interpenetrating (semi-IPN) networks of poly(vinylpyrrolidone)—PVP—and poly(2-hydroxyethyl methacrylate)—p(HEMA). Both polymers have been widely studied and used in medical and pharmaceutical fields. The idea was to combine the best features of both p(HEMA) and PVP hydrogels in a unique system with specific characteristics for the cleaning of works of art. In fact, PVP hydrogels exhibit high hydrophilicity but scarce mechanical features (Wang et al. 2010), whereas p(HEMA) hydrogels show good mechanical strength but low water-loading capa-

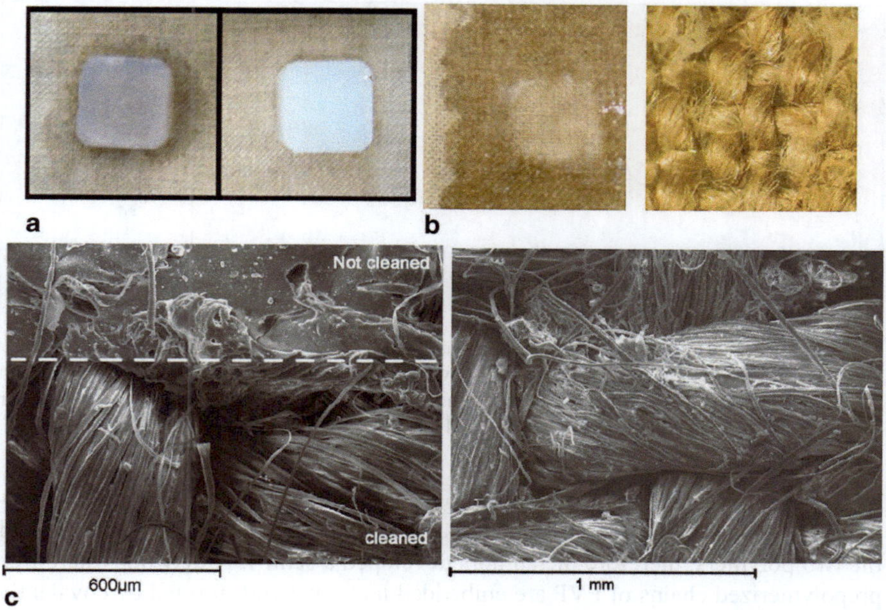

Fig. 4.12 **a** Two different formulations of acrylamide-based chemical hydrogels applied on canvas. The gel formulation can be tuned to improve water retention and avoid the diffusion of the loaded cleaning fluids across the surface of the canvas. **b** Removal of aged lining adhesives from canvas (on the right: optical microscope image, magnification × 10). **c** SEM images of the cleaned canvas. On the left: the border between the cleaned and not cleaned area is highlighted. (Readapted with permission from Pizzorusso et al. (2012))

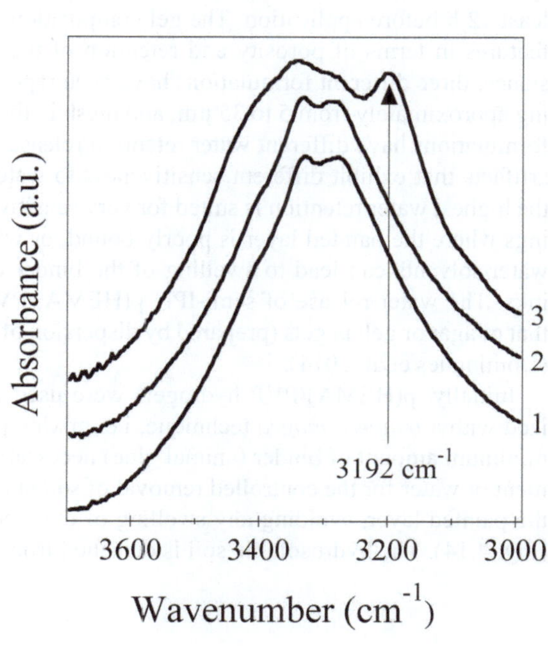

Fig. 4.13 FTIR-ATR spectra of: **(1)** canvas before the application of an acrylamide-based gel, **(2)** canvas after the application of the gel, **(3)** an acrylamide-based gel (the characteristic absorption band at 3192 cm^{-1} is due to the symmetric stretching of NH_2). The canvas spectra before and after application are very similar, and no bands typical of the acrylamide polymer are detected, confirming that acrylamide chemical gels do not leave detectable residues on the treated surfaces. (Reproduced with permission from Pizzorusso et al. (2012))

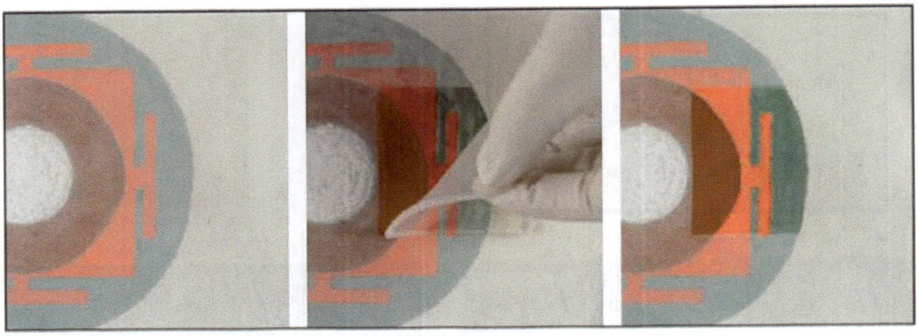

Fig. 4.14 Removal of grime from the surface of a Thang-Ka mock-up using a semi-IPN p(HEMA)/PVP chemical hydrogel. A Thang-Ka is a votive artifact based on *Tempera magra* painting on canvas, which typically exhibits a very low cohesive painted layer, with scarce quantity of water sensitive binder (e.g. animal glue). (Reproduced with permission from Domingues et al. (2013))

bility for application to cultural heritage conservation. Classical copolymer hydrogels may exhibit different characteristics from those resulting from the sum of the two polymers, therefore the authors developed a semi-IPN network where free prepolymerized chains of PVP are embedded into a network formed by covalently bonded HEMA monomers. These gels have been prepared in the shape of elastic foils that can be cut to desired size, manipulated and removed easily or "peeled" off from the surface (see Fig. 4.14) without leaving gel residues (as assessed through FTIR analysis), due to the cohesive network of chemical bonds. The gels are prepared as water containing polymer networks, but they can also be loaded with polar solvents (e.g. glycols, alcohols, ethanolamine) or water-based nanostructured systems (microemulsions, micellar solutions), typically through immersion for at least 12 h before application. The gel composition can be tuned to obtain different features in terms of porosity and retention of the loaded cleaning fluids. For instance, three different formulations have been reported, with macropores size ranging approximately from 5 to 35 μm, and mesh in the range of 2.5–3.1 nm. The three formulations have different water retention/release properties to allow treatment of artifacts that exhibit different sensitiveness to water or polar solvents. Obviously, the highest water retention is suited for very sensitive substrates such as easel paintings where the painted layer is poorly bound, or where the uncontrolled release of water/solvents can lead to swelling of the binder or leaching of colors, dyes and inks. The water release of semi-IPN p(HEMA)/PVP gels is more controlled than that of agar or gellan gels (prepared by dispersion of dry powders in water, 3 % w/w) (Domingues et al. 2014).

Initially, p(HEMA)/PVP hydrogels were used to clean canvas paintings realized with a *tempera magra* technique, i.e. mixing pigments and colorants with the minimum amount of binder (animal glue) necessary. The gels allowed the confinement of water for the controlled removal of soil embedded into the surface pores of the painted layer, avoiding any swelling of the binder or leaching of the pigments (Fig. 4.14). The hydrosoluble soil is detached from the surface and migrates inside

Fig. 4.15 Application of a semi-IPN p(HEMA)/PVP chemical gel loaded with a solvent blend for the removal of aged varnish (terpenic resin) from the surface of a canvas painting. The lower panel shows the removal of the varnish after the application of the gel (pictures under Vis and UV light), and the migration of the dissolved varnish inside the gel. (Photographs by Joana Domingues and Nicole Bonelli)

the gel. The p(HEMA)/PVP gels where also used for the removal of grime from the surface of inked paper samples, where the application of the highly retentive chemical hydrogel avoided any excessive wetting or alteration of the water-soluble ink (see also Chap. 1, Fig. 1.4). Moreover, the gels proved effective also for typical applications such as the removal of aged, yellowed terpenoid varnishes from the surface of canvas paintings (Figs. 4.15 and 4.16). In this case, the gels can be used to confine solvents or microemulsions and applied on the surface (contact time is usually 5–20 min). The varnish is either dissolved (and migrates inside the gel) or swollen (and then removed with a gentle mechanical action).

In fact, the versatility and effectiveness of these gels make them suitable for the cleaning of different works of art. In 2014, the best gel formulations developed at CSGI have been registered under the trademark Nanorestore gel®.

Beside chemical gels, researchers also developed "gel-like" dispersions of polymers with specific features for the cleaning of paintings. These formulations look like gels macroscopically, but cannot be strictly classified as gels from the

Fig. 4.16 Removal of aged varnish (terpenic resin) from the surface of a canvas painting, using a semi-IPN p(HEMA)/PVP chemical gel loaded with a solvent blend. The pictures (under Vis and UV light) show the final cleaning result on large portions of the painted layer. (Photographs by Joana Domingues and Nicole Bonelli)

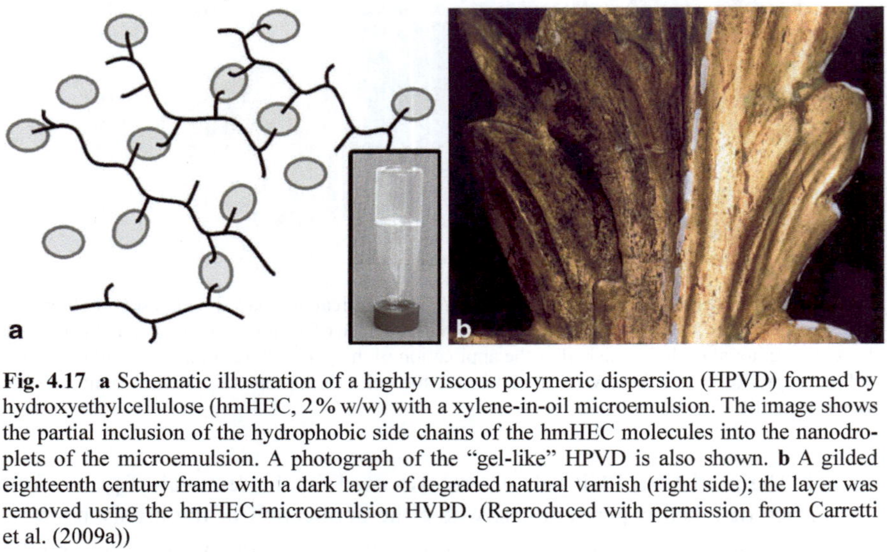

Fig. 4.17 **a** Schematic illustration of a highly viscous polymeric dispersion (HPVD) formed by hydroxyethylcellulose (hmHEC, 2 % w/w) with a xylene-in-oil microemulsion. The image shows the partial inclusion of the hydrophobic side chains of the hmHEC molecules into the nanodroplets of the microemulsion. A photograph of the "gel-like" HPVD is also shown. **b** A gilded eighteenth century frame with a dark layer of degraded natural varnish (right side); the layer was removed using the hmHEC-microemulsion HVPD. (Reproduced with permission from Carretti et al. (2009a))

rheological point of view; therefore they are referred to as Highly Viscous Polymeric Dispersions (HVPDs).

Carretti et al. (2009a) reported on the use of a HVPD of hydrophobically modified hydroxyethylcellulose (hmHEC, 2 % w/w) to embed a xylene-in-water microemulsion, for the cleaning of artifacts. Analysis through SAXS and DSC (Differential Scanning Calorimetry) indicated that the nanostructure of the microemulsion was retained upon loading inside the hmHEC network, with only minor modifications. In particular, the slight increase of the nanodroplet size and polydispersity, and of the average inter-droplet spacing, was attributed to a partial inclusion of the hydrophobic side chains of the hmHEC molecules into the nanodroplets (Fig. 4.17). Moreover, the zero-shear viscosity of the polymer dispersion increased significantly

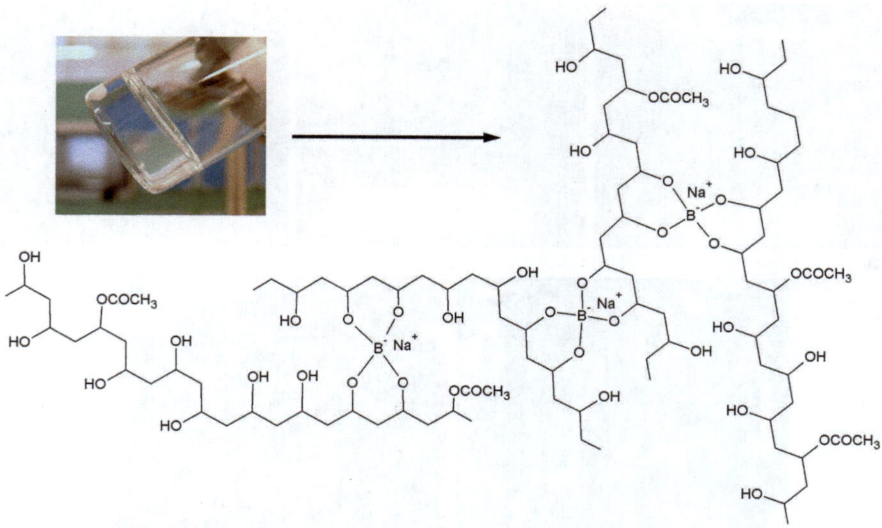

Fig. 4.18 Representation of a network formed by cross-links between partially hydrolyzed poly (vinyl acetate) and borax. (*Top Left*) An example of a PVAc-borax highly viscous polymeric dispersion. (Reproduced with permission from Natali et al. (2011))

upon loading the microemulsion, confirming the interaction between the polymer network and the nanostructured fluid. The hmHEC-microemulsion system was used to remove aged varnishes and coatings from the surface of paintings and gilded surfaces, as shown in Fig. 4.17.

Another possible approach is based on the use of HVPDs with high intrinsic elasticity, in order to achieve an easy removal of the gel-like system simply by a peeling action. Highly viscoelastic HVPDs can be obtained using borax and poly vinyl alcohol (PVA) or partially hydrolyzed poly vinyl acetate (PVAc). The gel-like network is due to the formation of cross-links between borax and hydroxyl groups of the PVA (or hydrolyzed PVAc), as shown in Fig. 4.18. The nature of the cross-links depends on pH, on temperature, on the concentration of the reagents, and on the chemical composition of the system (Koike et al. 1995; Wu et al. 1990; Keita et al. 1995).

Carretti et al. (2009b, 2010b) investigated the gelation properties of PVA-borax HVPDs containing water/organic liquid mixtures, for the cleaning of paintings. Due to their rheological properties, the PVA- and PVAc-borax HVPDs respond to the mechanical action involved in the removal of gels in a more elastic way than both traditional cellulose-based or polyacrylic acid-based hydrogels (see Fig. 4.19). In practice, this feature makes the removal of the "peelable" PVA-borax systems easier and safer, without leaving any residues on the treated surfaces as assessed by FTIR. Besides the easy removal, PVA-borax HVPDs allow the gradual and controlled cleaning of delicate, solvent-sensitive surfaces. For instance, a water-loaded formulation was recently used to remove a dark layer constituted by carbonaceous

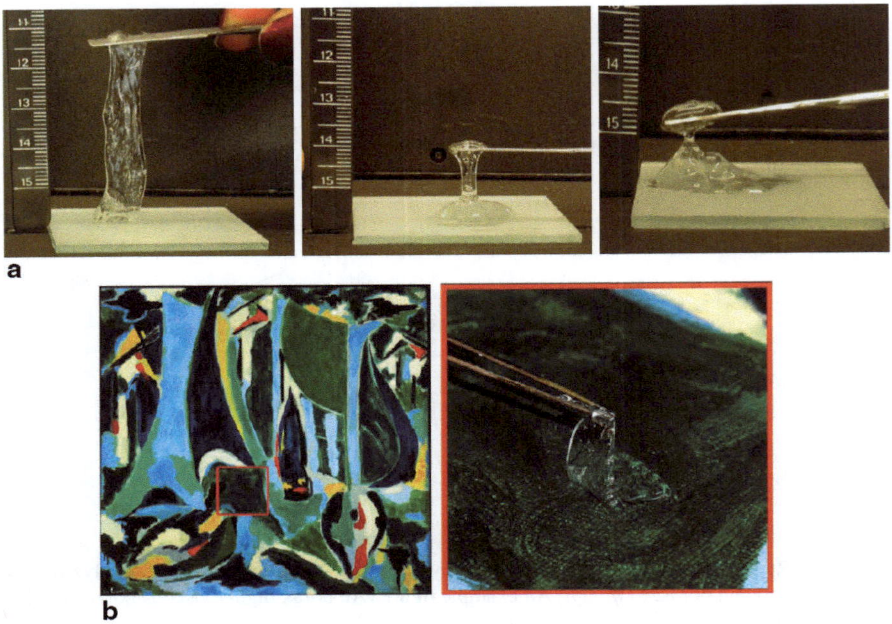

Fig. 4.19 **a** Removal by peeling of PVA-borax highly viscous polymeric dispersions with different PVA/borax content (*left, center*), and of a polyacrylic acid-based hydrogel (*right*). All the formulations contain 1-propanol (20% w/w). Reproduced with permission from "Poly(vinyl alcohol)—borate hydro/Cosolvent gels: viscoelastic properties, solubilizing powerm and application to art conservation. Langmuir 2009 25(15):8656–8662. Copyright 2009 American Chemical Society. **b** Photograph of the painting oil on canvas "Les Voiles" by Marcel Burtin (1902–1979) with the magnification of the region where the PVAc-borax polymeric dispersion was peeled off from the surface by means of tweezers. (Reproduced with permission from Natali et al. (2011))

particles (mainly due to atmospheric pollution) from the surface of a delicate painted layer that included water-sensitive substances like gypsum and white egg (as binder) (Carretti et al. 2014). In fact, the use of free, non-confined water led to solubilization of the dark layer but caused the swelling of the binder and local surface whitening, whereas the application of a water-loaded PVA-borax HVPD allowed the removal of the undesired layer without any damage to the painting.

It is possible to load different solvents in the PVA-borax formulations (up to ca. 15–30% w/w of the system liquid portion), including ethanol, 1-pentanol, 2-butanol, 1- and 2-propanol, acetone, cyclohexanone, *N*-methyl-pyrrolidinone and propylene carbonate. The use of partially hydrolyzed PVAc as gellant further extends the range of solvents that can be loaded, and allows the preparation of HVPDs where the organic solvent is up to 75% (w/w) of the liquid portion. It must also be noticed that the addition of an organic liquid to the HVPD leads to an increase in the number of cross-links (Angelova et al. 2011; Natali et al. 2011). Therefore, the selection of different gellants and loaded solvents influences the visco-elastic properties of the HVPD, and it is possible to adapt both the mechanical properties and the solubilizing power of these versatile systems in order to achieve selective and effective

Fig. 4.20 (*Left*) Wood panel by Ludovico Cardi (known as Cigoli, 1559–1613). The black box indicates the region where cleaning tests were carried out to remove old, oxidized varnishes. This picture was collected after the cleaning tests. (*Right, top*) Magnified view of the boxed region: (**a**) before application of the gel, (**b**) after two applications (4 min each) of a PVA-borax HPVD loaded with a water/1-propanol mixture. The red dashed lines indicate the area where the gel was applied. Reproduced with permission from Poly(vinyl alcohol)—borate hydro/Cosolvent gels: viscoelastic properties, solubilizing powerm and application to art conservation. Langmuir 2009 25(15):8656–8662. Copyright 2009 American Chemical Society. (*Right, bottom*) Region where three different PVAc-borax HVPDs (loaded with water/1-propanol mixtures)were applied to achieve different degrees of varnish removal. The conservators who conducted the cleaning tests deemed the formulations used on areas "B" and "C" excellent, whereas the system used on area "A" led to a too fast removal of the varnish. Reproduced with permission from Natali et al. (2011)

cleaning results on different artistic/historical substrates. Figure 4.20 shows the application of different PVA- and PVAc-borax HVPDs loaded with water/1-propanol mixtures for the removal of oxidized varnishes from the surface of a late Renaissance wood panel. In general, depending on the solvent used, the varnishes can be either dissolved (migrating into the gel-like system) or swollen and detached from the surface (in this case the removal is completed by gentle mechanical action with a swab) (Fig. 4.20).

In the following sections, both the application field and practical procedures for using the most effective gel and gel-like formulations will be detailed. As mentioned in the previous chapters, it is important to recall that each case study exhibits specific characteristics; therefore the practical guidelines should be considered and applied critically, and not merely as recipes. Moreover, altering the composition of the cleaning systems can change their structure and effectiveness. Therefore, the best approach is always the collaboration and interactive feedback between the conservation scientists who designed the cleaning systems and conservators, to grant the flexibility and adaptability of the methods, minimizing (or avoiding) drawbacks and maximizing the results.

4.3 Application Field

As mentioned in Sect. 4.1, the use of gels in cultural heritage preservation is strictly connected to the need of controlling liquid systems for the cleaning of surfaces. In fact, organic solvents with low surface tension quickly soak the substrate through capillarity and diffusion processes. This is particularly dangerous on easel paintings where a complex stratigraphy of painted layers is laid on supports such as fabric or wood, which can load a large amount of solvents and expand or shrink anisotropically because of the adsorption/desorption processes. The final effect is the detachment of the paint layer. At the same time, the partial swelling or solubilization of original organic components of the paint layer (e.g. binder media) and of the preparation layer (i.e. glue) may occur. The confinement of solvents into gels allows control of the cleaning intervention: the spreading of liquids within the substrate is limited and the cleaning action is focused on the surface of the work of art. Moreover, confined/gelled liquids can be easily handled and adapted to the morphology and spatial constraints of the artifact.

The use of gels in conservation, as already suggested, started on easel paintings; thereafter, these tools have been used on numerous artistic and historical substrates such as paper, parchment, metals, ceramics, and others. Beside the control of the cleaning action, the main advantages include low environmental impact, feasible application and the possibility of checking in real time the cleaning process due to the transparency of the gel.

The versatility of this technique depends on the specific characteristics of the different types of gels. Classic thickening agents, such as polyacrylic acid, have different application fields than high-retention chemical gels. Different rheological properties, in terms of elasticity and viscosity, involve different application fields and applicative procedures.

For instance, as described in the previous sections the preparation of gel-like dispersions using thickening agents is relatively simple and not time consuming; however, these "gels" exhibit a limited retention capability and are often prone to leaving residues that are removed with solvents or aqueous solutions, therefore these tools are not recommended on solvent-sensitive works of art, e.g. delicate or poorly bound painted layers, tapestry, artworks with soluble inks/dyes etc. In fact, in most cases a higher control of solvent diffusion is required, and it is preferable to use highly retentive chemical gels, which work as 'sponges' able to load solvents and release them at controlled rate. The high retention capability of these systems permits limiting the cleaning action to the contact area of the gel with the artifact surface. This property largely expands the application field of this class of gels. Recently, these tools have been tested for the removal of surface dirt or aged/detrimental adhesives and coatings from the surface of canvas paintings (including *tempera magra*, i.e. poorly bound paintings), drawings, manuscripts, parchment, leather, and silk tissues. Obviously, the controlled release of the confined cleaning liquids might involve longer application time than traditional physical gels, thickened solvents and cellulose poultices loaded with liquids (see Chap. 3), but the high retentiveness fully compensate for this possible concern. In fact, excessive wetting or spreading can be avoided even during long applications (see below and next section).

As anticipated in Sect. 4.2, the water release values of three different formulations of p(HEMA)/PVP chemical hydrogels (named "H50", "H58" and "H65") on a highly sorbent hydrophilic substrate (filter paper) were compared to those of agar and gellan-based gels (prepared by dispersion of dry powders in water, 3 % w/w) (Domingues et al. 2014). Results showed that the water release of p(HEMA)/PVP gels was either ca. one fourth (8 mg/cm^2, H50) or one half (15–16 mg/cm^2, H58–H65) of agar or gellan formulations' (30–33 mg/cm^2). In Chap. 1 we illustrated an example where the highly retentive chemical hydrogels allowed cleaning of surfaces without altering water-sensitive colors, indicating that the use of p(HEMA)/PVP gels avoided excessive wetting (see Fig. 1.4). Indeed, chemical hydrogels, simply loaded with water or water-ethanol blends, have also been used for the controlled removal of soil from the surface of inked parchment, which exhibits high sensitiveness to traditional cleaning interventions.

The application field can be further expanded when gels are used to confine nanostructured cleaning fluids such as microemulsions and micellar solutions, which have a reduced health and environmental impact as compared to pure solvents, and are highly effective in the removal of detrimental synthetic coatings (Giorgi et al. 2010; Baglioni 2012a, 2012b). For instance, p(HEMA)/PVP hydrogels were loaded with the water-based EAPC fluid (described in Sect. 4.2 and in Chap. 3), for the removal of an aged ethyl acrylate/methyl methacrylate adhesive (Plextol® B500) from canvas. The controlled release of EAPC allowed the gradual swelling and softening of the synthetic lining adhesive, which was then removed with a gentle mechanical action without damaging the canvas fibers. Moreover, excessive wetting was avoided despite the long application time (4 h) (Domingues et al. 2014). The p(HEMA)/PVP hydrogels were also used for the cleaning of seventeenth century terrestrial globes by Vincenzo Coronelli: water-loaded gels were useful for the removal of hydrosoluble surface dirt, while EAPC-loaded gels were used for the non-invasive and fast swelling and detachment of aged PVAc-based coatings from the surface of the globes (contact time ca. 10 min).

4.4 Application Procedures

The aim of this section is to provide practical guidelines for the application of peelable and highly retentive hydrogels on easel paintings. In particular, these guidelines are to be used for the application of gels as designed at CSGI (Consorzio Interuniversitario per lo Sviluppo dei Sistemi a Grande Interfase, Center for Colloid and Surface Science)—University of Florence, and registered as Nanorestore Gel®. Different formulations might behave differently, so adaptations of the indications here provided should be considered case by case.

The proposed methods are based on the theoretical and practical knowledge gathered in the last decade through both research work and field activity in collaboration with conservators. It is fundamental to recall that conservation issues can vary greatly; therefore the practical guidelines reported here should be considered and applied critically, and not as recipes. Conservators are encouraged to contact the authors of this book for further guidance and recommendations on the use of these formulations.

In particular, in this section we chose to report on the applicative procedures for using semi-IPN p(HEMA)/PVP chemical hydrogels (which will be referred to as "p(HEMA)/PVP gels") and PVA-Borax "gel-like" dispersions, as these proved to be highly versatile and effective systems for the removal of grime and undesired layers from the surface of works of art.

p(HEMA)/PVP gels are typically produced as water-loaded, transparent elastic foils (for instance 10×20 cm, 2 mm thick, but the size can be changed according to different requirements, see Figs. 4.14, 4.15 and 4.16). When packaged, the foils are immersed in a small amount of (purified) water. During storage, they must be kept immersed in water, for instance in sealed vials or containers. Under these conditions, the gels are very stable, and can be stored at room temperature for long periods (e.g. more than 1 year) before use. However, as a general rule, it is advisable to check gels after long storage periods, to verify that no alteration has occurred such as discoloration, texture change, opacity, rupture etc. The gels are fairly resistant to microbial attack but as a general criterion the water in the containers can be replaced periodically throughout long storage periods to prevent contamination from microorganisms.

p(HEMA)/PVP gels are non-toxic, and can be handled using standard laboratory gloves according to good laboratory practice. For solvent-containing gels, all the standard precautions related to the use of solvents must be taken, even though the volatility (and consequently the toxicity) of solvents is reduced when they are confined into gels.

Water-loaded p(HEMA)/PVP gels can be used directly for the removal of hydrosoluble dirt from the surface of paintings. The gels are removed from their container (they are kept immersed in a small amount of water, see above), cut down to desired size if need be, and then the water excess on the gel surface is gently removed by placing each face of the gel on a small sheet of filter paper for a very short time (ca. 1 s). Then, the gel is applied on the artifact surface, for instance handling it or using tweezers. These gels have been used both on horizontal and vertical surfaces, and even on ceilings, however adherence is influenced by several factors, such as the physico–chemical nature of the surface (including adhesive coatings, varnishes etc.), the type of gel used (more or less elastic), and the cleaning fluid loaded into the gel. A transparent Mylar® film can be placed over the gel to prevent evaporation of water from the gel network during the application, even if for standard cleaning operations with water-loaded gels this is not strictly necessary as the contact time might be short enough (as compared to the evaporation rate of water confined in the gel) to keep the gel wet during the whole cleaning intervention. In fact, contact time is strongly depending on several factors, including the nature of the surface and of the dirt/grime that needs to be removed, and the type of gel used. In practical applications contact time can vary from a minute up to 3–4 h, even if short applications (1–20 min) have been reported more frequently. However, it is important to avoid drying of the gel during the intervention (the gel must remain humid). The removal of the gel can be carried out easily using tweezers or simply by handling it ("peeling" the gel). The application of water-loaded gels can lead either to solubilization or swelling/detachment of hydrosoluble dirt/grime layers. In the first case, migration of the dissolved dirt into the gel network can be directly observed (e.g. the gel yellows or darkens). If the dirt layer is swollen, after gel removal a gentle mechanical action (e.g. using a swab) can be carried out to remove the swollen and softened dirt. The gel can also be applied in short successive applications (rather than a single, continuous application), each time checking the solubilization or detachment of the dirt layer and re-applying the gel if need be (gels must not dry, as explained above). After the final cleaning result has been achieved, the gel can be immersed in water (e.g. overnight or for ca. 12 h) to release dirt and re-load pure water, and then used again or stored in pure water for reuse in the short term. Long-term storage of gels following the first application is not advisable, and the gels conditions should be checked periodically. The number of times that a single gel can be re-loaded with pure water and reused depends on the physico–chemical nature and amount of material that has been dissolved and carried into the gel. For instance, it has been reported that for dust/dirt removal, water-loaded gels have been reused roughly up to 3 times, but the cleaning effectiveness might decrease progressively also depending on the nature of the dirt/grime. In case the dirt/grime layer was swollen and detached, but did not migrate into the gel network, it might be possible to reuse the gel more times.

Loading pHEMA/PVP gels with solvents (e.g. glycols, alcohols, ethanolamine), water-solvent blends (e.g. water and ethanol) or microemulsions can be carried out by immersing the original water-loaded gel into the desired cleaning fluid for at least 12 h. The application is similar as for water-loaded gels, however in this case it is important to place a covering film over the gel during application, in order to avoid evaporation of the loaded cleaning fluid (solvents or microemulsions). Obviously, the covering film (e.g. Mylar® or other) must not be soluble in the cleaning fluid. Migration of terpenoid varnishes such as dammar into pHEMA/PVP gels loaded with solvents has been reported, while for synthetic coatings/varnishes (e.g. poly vinyl acetate-based adhesives) the swelling and detachment of the coating has also been observed; the swollen/detached coating is then removed with a gentle mechanical action, for instance using a swab. A final non-invasive clearance step can be carried out, when necessary, using a pHEMA/PVP gel simply loaded with water, to remove water-soluble non-volatile components (e.g. non-degradable surfactants, soaps, etc. which in some cases might remain as a residue over the cleaned surface) present in the cleaning fluids that were loaded in the gels. The shelf life of the EAPC microemulsion is ca. 2–3 months. After this time the ethyl acetate present in the microemulsion slowly degrades and although the microemulsion can still be used for cleaning, a post-cleaning treatment with $Ca(OH)_2$ nanoparticles (see Sect. 2.4.2) is recommended.

Gels might be re-loaded in fresh solvents/microemulsions and reused (few times), but each time it is important to check that no alteration of the gel network has occurred (discoloration, rupture, etc.). In fact, the interaction of the removed coating with the gel network depends on the physico–chemical nature and amount of the removed material.

The main procedures for the application of pHEMA-PVP gels are illustrated in Figs. 4.21, 4.22.

PVA-Borax "gel-like" dispersions are directly prepared as loaded with water or with water/organic solvent blends. The dispersions can be handled using standard laboratory gloves. For solvent-loaded formulations, all the standard precautions related to the use of solvents must be taken, even though the volatility (and consequently the toxicity) of solvents is reduced when they are confined into gels. Moreover, the solvents are present as water/organic blends as explained in Sect. 4.2. The dispersions are kept in sealed vials (to prevent evaporation of solvents), at room temperature. Storage periods as long as 1 year have been reported. However, as a general rule, it is advisable to check gels after long storage to verify that no alteration has occurred such as discoloration/opacity, rupture etc.

The transparent, highly viscoelastic dispersions are applied directly onto the painted surface and modeled with a spatula to cover the desired area. Portions of gels can also be preliminary modeled and then gently applied on the surface using a spatula. These systems show optimal adherence to the surface (they have been used even on vertical surfaces) and adapt their shape to the surface roughness. Contact time is normally of some minutes. Short successive applications (e.g. 1–4 min each) are used to gradually remove dirt or unwanted layers from painted surfaces. Dirt/coatings are either dissolved (and migrate inside the gel) or detached/swollen. When the desired degree of cleaning has been achieved, the PVA-borax dispersion is simply removed ("peeled") using tweezers. In some cases, as mentioned in Sect. 4.2, a gentle mechanical action (for instance using a swab) is needed to remove the swollen grime or coatings from the surface of the artifact.

The main procedure for the application of PVA-Borax HVPDs (Highly Viscous Polymeric Dispersions) is illustrated in Fig. 4.23.

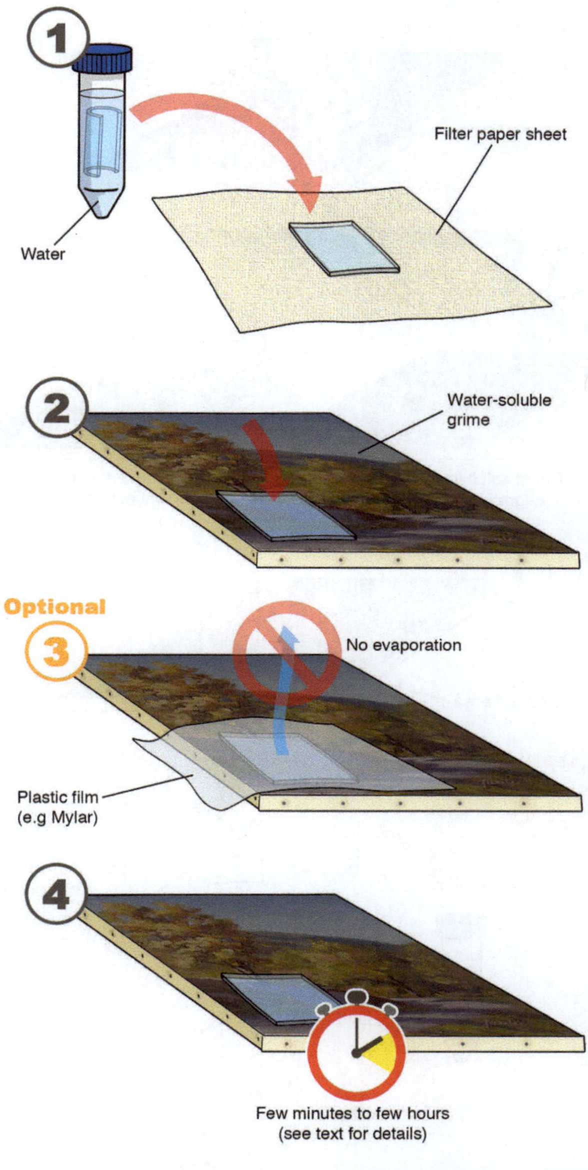

Fig. 4.21a The main practical steps for the application of pHEMA-PVP gels (loaded with water) for the removal of hydrosoluble dirt from the surface of easel paintings. Part I. *1* The gel is removed from its container (pHEMA-PVP gels are normally kept immersed in small amounts of water). The water excess on the gel surface is gently removed by placing each face of the gel on a small sheet of filter paper for a very short time (ca. 1 s). *2* The gel is applied on the artifact surface. *3* A transparent Mylar® film can be placed over the gel to prevent evaporation of water from the gel network during the application. This is not strictly necessary as the contact time might be short enough to keep the gel wet during the whole cleaning intervention. *4* Contact time can vary from a minute up to 3-4 h, even if short applications (1-20 min) have been reported more frequently. It is important to avoid drying of the gel during the intervention (the gel must remain humid). (Image by Michele Baglioni)

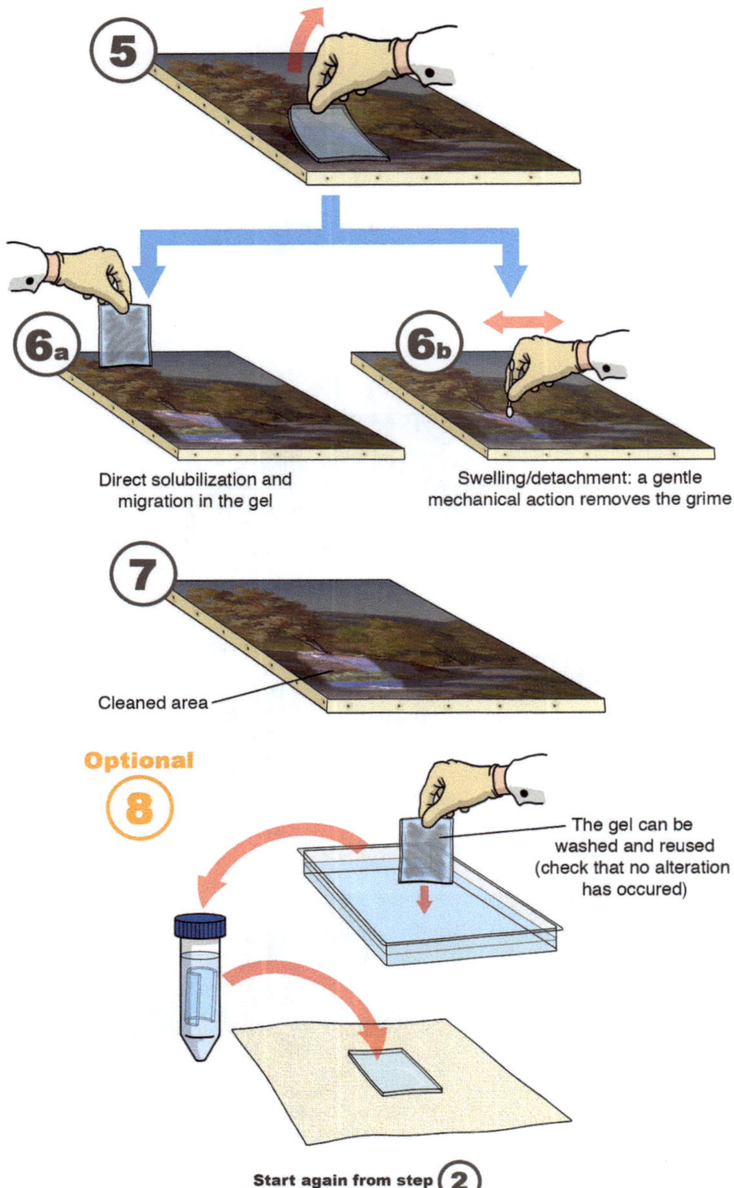

Fig. 4.21b The main practical steps for the application of pHEMA-PVP gels (loaded with water) for the removal of hydrosoluble dirt from the surface of easel paintings. Part II. *5* Removal of the gel. *6* Dirt/grime is usually either dissolved (and it migrates into the gel as shown in panel *6a*), or swollen/detached (and then removed with a gentle mechanical action after removal of the gel, as shown in panel *6b*). *7* The cleaned painting (following *6a* or *6b*). *8* The used gel can be immersed in water (e.g. overnight or for ca. 12 h) to release dirt and re-load pure water, and then stored in pure water or used again. Long-term storage of gels following the first application is not advisable, and the gels conditions should be checked periodically (see text for details). (Image by Michele Baglioni)

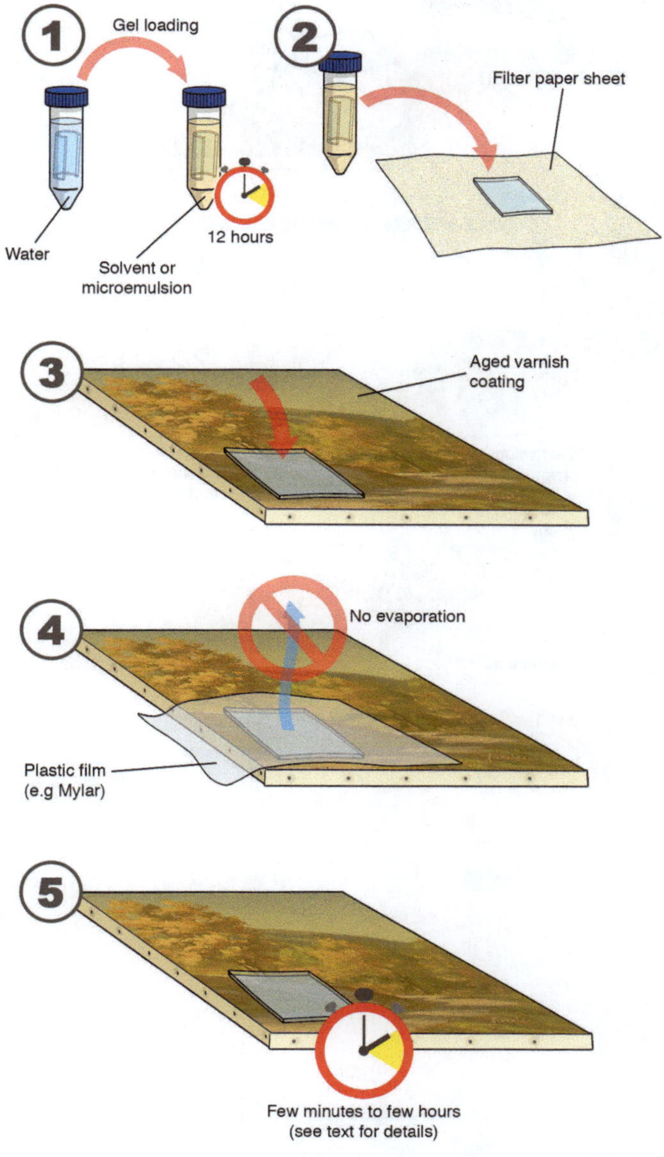

Fig. 4.22a The main practical steps for the application of pHEMA-PVP gels (loaded with solvents or microemulsions/micellar solutions) for the removal of aged varnishes/coatings from the surface of easel paintings. Part I. *1* The gels are loaded with solvents or microemulsions by immersing the original water-loaded gel into the desired cleaning liquid for at least 12 h. *2* The gel is removed from its container and the liquid excess on the gel surface is gently removed by placing each face of the gel on a small sheet of filter paper for a very short time (ca. 1 s). *3* The gel is applied on the artifact surface. *4* A transparent covering film (e.g. Mylar®) is placed over the gel to prevent evaporation of the cleaning liquid from the gel network during the application. Obviously, the covering film must not be soluble in the cleaning liquid. *5* Contact time can vary from a minute up to 3-4 h, even if short applications (1–20 min) have been reported more frequently. It is important to avoid drying of the gel during the intervention (the gel must remain humid). (Image by Michele Baglioni)

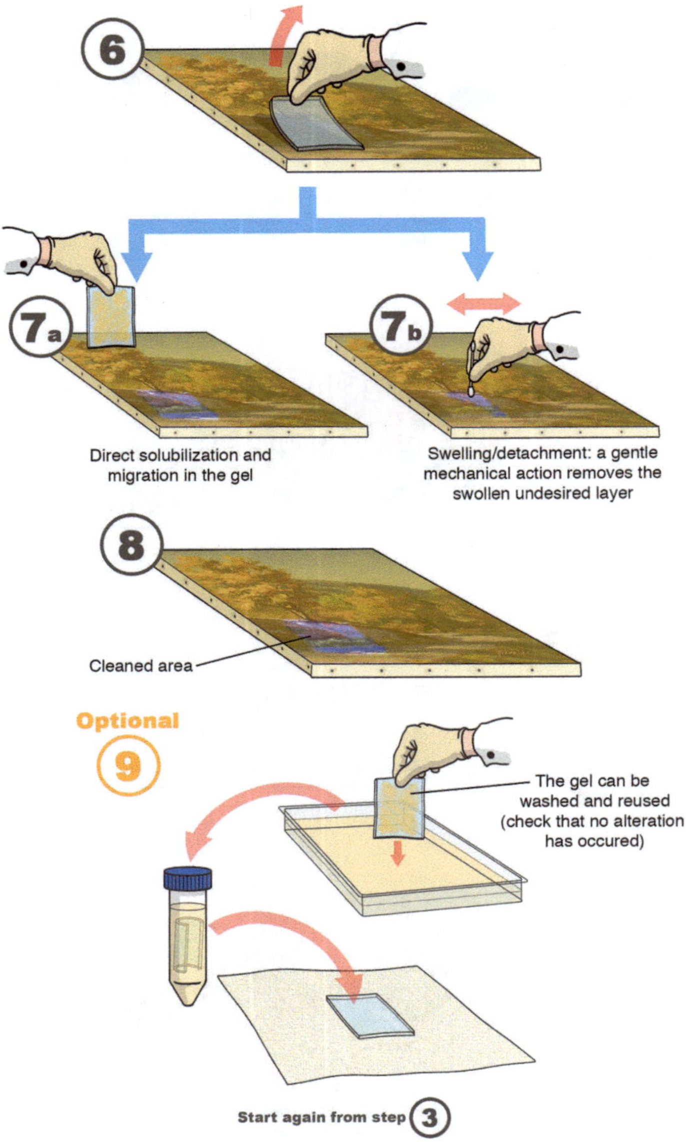

Fig. 4.22b The main practical steps for the application of pHEMA-PVP gels (loaded with solvents or microemulsions/micellar solutions) for the removal of aged varnishes/coatings from the surface of easel paintings. Part II. *6* Removal of the gel. *7* The varnish/coating is usually either dissolved (and it migrates into the gel as shown in panel *7a*, or swollen/detached (and then removed with a gentle mechanical action after removal of the gel, as shown in panel *7b*). *8* The cleaned painting (following *7a* or *7b*). *9* The used gel can be immersed in the cleaning liquid (solvents, microemulsions/micellar solutions) overnight or for ca. 12 h, to release entrapped matter and re-load fresh cleaning liquid, and then stored or used again. Long-term storage of gels following the first application is not advisable, and the gels conditions should be checked periodically (see text for details). (Image by Michele Baglioni)

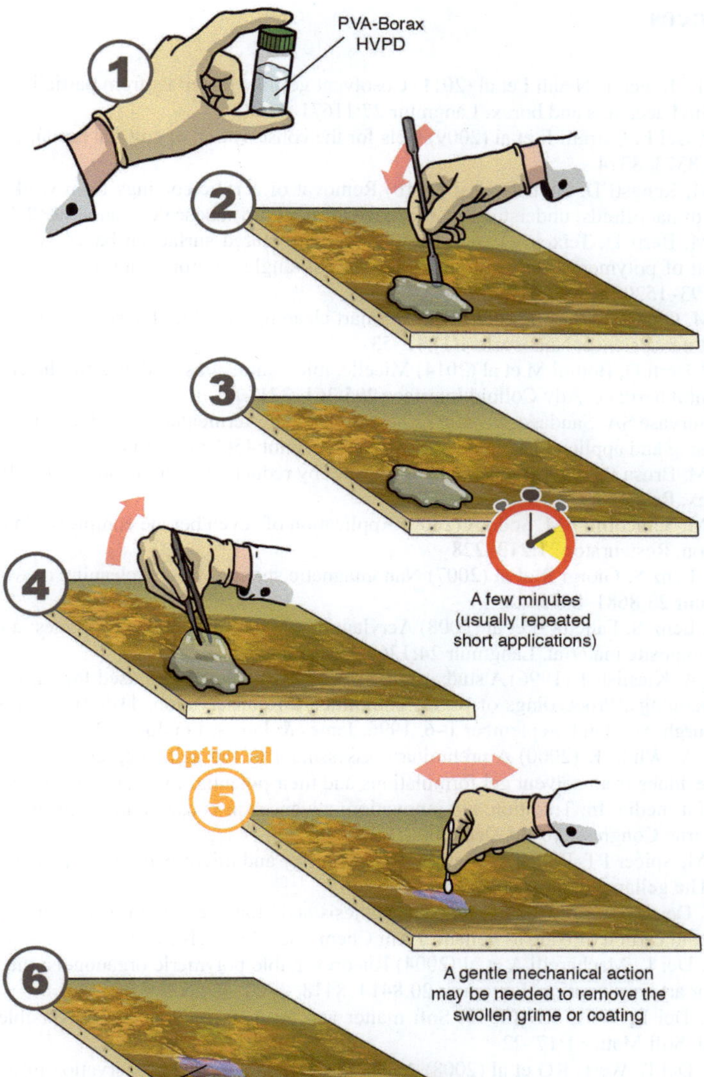

Fig. 4.23 The main practical steps for the application of PVA-Borax HVPDs (Highly Viscous Polymeric Dispersions) for the removal of grime or varnishes/coatings from the surface of easel paintings. *1* The HVPDs are kept in sealed vials (to prevent evaporation of solvents), at room temperature. *2* The transparent HVPDs are applied directly onto the painted surface and modeled with a spatula to cover the desired area. Portions of gels can also be preliminary modeled and then gently applied on the surface using a spatula. *3* Contact time is normally of some minutes. Short successive applications (e.g. 1–4 min each) are used to gradually remove dirt or unwanted layers from painted surfaces. *4* Grime or varnishes/coatings are either dissolved (and migrate inside the gel) or detached/swollen. When the desired degree of cleaning has been achieved, the PVA-borax dispersion is simply removed ("peeled") using tweezers. *5* In some cases a gentle mechanical action (for instance using a swab) is needed to remove the swollen grime or coatings from the surface of the artifact. *6* The cleaned painting (following step *4* or *5*). (Image by Michele Baglioni)

References

Angelova L, Terech P, Natali I et al (2011) Cosolvent gel-like materials from partially hydrolyzed poly(vinyl acetate)s and borax. Langmuir 27:11671–11682

Baglioni P, Dei L, Carretti E et al (2009) Gels for the conservation of cultural heritage. Langmuir 25(15):8373–8374

Baglioni M, Rengstl D, Berti D et al (2010) Removal of acrylic coatings from works of art by means of nanofluids: understanding the mechanism at the nanoscale. Nanoscale 2:1723–1732

Baglioni M, Berti D, Teixeira J et al (2012a) Nanostructured surfactant-based systems for the removal of polymers from wall paintings: a small-angle neutron scattering study. Langmuir 28:15193–15202

Baglioni M, Giorgi R, Berti D et al (2012b) Smart cleaning of cultural heritage: a new challenge for soft nanoscience. Nanoscale 4(1):42–53

Baglioni P, Berti D, Bonini M et al (2014) Micelle, microemulsions, and gels for the conservation of cultural heritage. Adv Colloid Interface 205:361–371

Bajaj IB, Survase SA, Saudagar PS et al (2007) Gellan gum: fermentative production, downstream processing and applications. Food Technolo Biotechnol 45(4):341–354

Bicchieri M, Brusa P (1997) The bleaching of paper by reduction with the borane *tert*-butylamine complex. Restaurator 18:1–11

Bicchieri M, Sementilli FM, Sodo A (2000) Application of seven borane complexes in paper conservation. Restaurator 21:213–228

Bonini M, Lenz S, Giorgi R et al (2007) Nanomagnetic sponges for the cleaning of works of art. Langmuir 23:8681–8685

Bonini M, Lenz S, Falletta E et al (2008) Acrylamide-based magnetic nanosponges: a new smart nanocomposite material. Langmuir 24:12644–12650

Burnstock A, Kieslish T (1996) A study of the clearance of solvent gels used for varnish removal from paintings. Proceedings of ICOM committee for conservation, 11th Triennal meeting in Edimburgh, Scotland, September 1–6, 1996. James & James, London, p 253

Burnstock A, White R (2000) A preliminary assessment of the aging/degradation of Ethomeen C-12 residues from solvent gel formulations and their potential for inducing changes in resinous paint media. In: Tradition and innovation: advances in conservation, contributions to the Melbourne Congress, 10–14 Oct 2000 (IIC), p 38

Caggioni M, Spicer PT, Blair DL et al. (2007) Rheology and microrheology of a microstructured fluid: The gellan gum case. J Rheol 51:851–865

Carretti E, Dei L, Baglioni P et al (2003) Synthesis and characterization of gels from polyallylamine and carbon dioxide as gellant. J Am Chem Soc 125:5121–5129

Carretti E, Dei L, Macherelli A et al (2004) Rheoreversible polymeric organogels: the art of science for art conservation. Langmuir 20:8414–8418

Carretti E, Dei L, Weiss RG (2005) Soft matter and art conservation. Rheoreversible gels and beyond. Soft Matter 1:17–22

Carretti E, Dei L, Weiss RG et al (2008) A new class of gels for the conservation of painted surfaces. J Cult Herit 9:386–393

Carretti E, Fratini E, Berti D et al (2009a) Nanoscience for art conservation: oil-in-water microemulsions embedded in a polymeric network for the cleaning of works of art. Angew Chem Int Ed Engl 48:8966–8969

Carretti E, Grassi S, Cossalter M et al (2009b) Poly(vinyl alcohol)-Borate Hydro/Cosolvent Gels:Viscoelastic Properties, Solubilizing Power, and Application to Art Conservation. Langmuir 25(15):8656–8662

Carretti E, Bonini M, Dei L et al (2010a) New frontiers in materials science for art conservation: responsive gels and beyond. Accounts Chem Res 43(6):751–760

Carretti E, Natali I, Matarrese C et al (2010b) A new family of high viscosity polymeric dispersions for cleaning easel paintings. J Cult Herit 11:373–380

Carretti E, Matarrese C, Fratini E et al (2014) Physicochemical characterization of partially hydrolyzed poly(vinyl acetate)-borate aqueous dispersions. Soft Matter 10:4443–4450

Casoli A, Di Diego Z, Isca C (2014) Cleaning painted surfaces: evaluation of leaching phenomenon induced by solvents applied for the removal of gel residues. Environ Sci Pollut Res. doi:10.1007/s11356-014-2658-5

Chelazzi D, Chevalier A, Pizzorusso G et al (2014) Characterization and degradation of poly(vinyl acetate)-based adhesives for canvas paintings. Polym Degrad Stab 107:314–320. http://dx.doi.org/10.1016/j.polymdegradstab.2013.12.028

Chevalier A, Chelazzi D, Baglioni P et al (2008) Extraction d'adhésifs de rentoilage en peinture de chevalet: nouvelle approche. Proceedings of the ICOM-CC 15th Triennal Conference, 22–26 September 2008, vol 2. New Dehli (India), 2008, p 581

Cremonesi P (2006) Applicazione di metodologie di intervento più recenti per la pulitura del material cartaceo. In Atti delle giornate di studio, Problemi di Restauro - Giornate di Studio per storici d'arte, Ispettori di Soprintendenza, conservatori e operatori museali. Il Prato, p 39

Cremonesi P, Signorini E (2012) Un approccio alla pulitura dei dipinti mobili, I Talenti - Metodologie, tecniche e formazione nel mondo del restauro, 29, Casa Editrice Il Prato, Saonara (PD, Italy)

Davies E, Huang Y, Harper JB et al (2010) Dynamics of water in agar gels studied using low and high resolution ^{1}H NMR spectroscopy. Int J Food Sci Tech 45(12):2502–2507

Domingues J, Bonelli N, Giorgi R et al (2013) Innovative hydrogels based on semi-interpenetrating p(HEMA)/PVP networks for the cleaning of water-sensitive cultural heritage artifacts. Langmuir 29:2746–2755

Domingues J, Bonelli N, Giorgi R et al (2014) Chemical semi-IPN hydrogels for the removal of adhesives from canvas paintings. Appl Phys A 114:705–710

Giorgi R, Michele B, Berti D et al (2010) New methodologies for the conservation of cultural heritage: micellar solutions, microemulsions, and hydroxide nanoparticles. Accounts Chem Res 43(6):695–704

Gorel F (2010) Assessment of agar gel loaded with micro-emulsion for the cleaning of porous surfaces. CeROArt: conservation, exposition, restauration d'objets d'art. http://ceroart.revues.org/1827. Accessed 6 June 2014

Gulotta D, Saviello D, Gherardi F et al (2014) Setup of a sustainable indoor cleaning methodology for the sculpted stone surfaces of the Duomo of Milan. Herit Sci 2:6–19

Iannucelli S, Sotgiu S (2010) Wet treatments of works of art on paper with rigid gellan gels. Book Pap Group Ann 29:25–39

Keita H, Ricard A, Audebert R et al (1995) The poly(vinyl alcohol)-borate system: influence of polyelectrolyte effects on phase diagrams. Polymer 36(1):49–54

Koike A, Nemoto N, Inoue T et al (1995) Dynamic light scattering and dynamic viscoelasticity of poly(vinyl alcohol) in aqueous borax solutions. 1. Concentration effect. Macromolecules 28:2339–2344

Mao R, Tang J, Swanson BG (2001) Water holding capacity and microstructure of gellan gels. Carbohydr Polym 46:365–371

Nakamura K, Shinoda E, Tokita M (2001) The influence of compression velocity on strength and structure for gellan gels. Food Hydrocolloids 15:247–252

Natali I, Carretti E, Angelova L et al (2011) Structural and mechanical properties of "peelable" organoaqueous dispersions with partially hydrolyzed poly(vinyl acetate)-borate networks: applications to cleaning painted surfaces. Langmuir 27:13226–13235

Pizzorusso G, Fratini E, Eiblmeier J et al (2012) Physicochemical characterization of acrylamide/bisacrylamide hydrogels and their application for the conservation of easel paintings. Langmuir 28:3952–3961

Stulik D et al (2004) Solvent gels for the cleaning of works of art: the residue question. Dorge V (ed). Getty Publications, Los Angeles

Wang J, Sun F, Li X (2010) Preparation and antidehydration of interpenetrating polymer network hydrogels based on 2-hydroxyethyl methacrylate and *n*-vinyl-2-pyrrolidone. J Appl Polym Sci 117:1851–1858

Wolbers R, Sterman N, Stavroudis C (1988) Notes for the Workshop on New Methods in Cleaning of Paintings. The Getty Conservation Institute, Marina del Rey, CA, 1988
Wolbers R (2000) Cleaning painted surfaces: aqueous methods. Archetype, London
Wu WL, Shibayama M, Roy S et al (1990) Physical gels of aqueous poly(vinyl alcohol) solutions: a small-angle neutron-scattering study. Macromolecules 23(8):2245–2251

Chapter 5
Deacidification of Paper, Canvas and Wood

Abstract Cellulose-based materials are affected by acid hydrolysis and oxidation processes, which eventually cause aesthetic alteration and the loss of artifacts' mechanical properties. Numerous methods have been proposed and applied in the last 50 years to counteract cellulose degradation. Aqueous treatments involve drawbacks such as leaching of inks and the swelling of cellulose fibers. Several non-aqueous methods use precursors for the in situ formation of $Mg(OH)_2$, which eventually forms carbonate as a buffer against recurring acidity. Anti-oxidants have been coupled to deacidifying agents to complete the treatment. Alternatively, it has been shown that nanoparticles of $Ca(OH)_2$ or $Mg(OH)_2$ can be dispersed in different solvents (e.g. short chain alcohols, cyclohexane) and applied on artifacts, to stably neutralize acidity and inhibit the oxidative activity of iron and copper ions, granting long-term protection.

This chapter illustrates the main formulations of nanoparticles that have been developed in the last decades for the deacidification of paper, canvas and wood, providing references to real case studies and application fields. Finally, a whole section is dedicated to applicative procedures and protocols for the use of nanoparticles dispersions.

5.1 Background

The preservation of cellulose-based objects is a central topic in science for the conservation of cultural heritage. In fact, cellulose is the main constituent of paper and canvas, and one of the main components of wood; therefore the prevention, limitation or counteraction of cellulose degradation is beneficial for a large variety of intellectual, historical and artistic heritage, such as canvas paintings, textiles, books, manuscripts, maps, drawings and documents in archives or libraries, as well as wooden objects ranging from sculptures to shipwrecks. This chapter reports on the use of nanoparticles dispersions for the pH control of movable works of art, namely paper, wood and canvas objects, discussing the characteristics and the practical aspects of this approach.

P. Baglioni et al., *Nanotechnologies in the Conservation of Cultural Heritage*,
DOI 10.1007/978-94-017-9303-2_5

A thorough description of the macroscopic and microscopic structure and of the chemistry of all the components that might be included in cellulose-based artifacts (e.g. besides cellulose: hemicellulose, lignin, pectin, extractives, etc.) is beyond the scope of this book, and the interested reader can refer to a vast amount of literature on such topics. However, it is important to recall the fundamental notions that are useful for discussing the conservation issues and the role of the nanomaterials used to counteract degradation processes.

Cellulose is the major constituent of paper, cotton, linen and hemp; moreover it is also the main component of jute (ca. 65 %, plus ca. 22 % of hemicellulose) and of wood (roughly 40–45 %, plus ca. 15–25 % of hemicellulose). Cellulose is a natural linear polysaccharide produced by plants, and it consists of D-glucose units (pyranose rings) that are linked each other through β-(1,4)-glycosidic bonds. The characteristic 180° rotation around the 1–4 axis between two following glucose units reduces steric encumbrance, and formally the repeating unit (cellobiose) is made of two glucose molecules (see Fig. 5.1a).

The degree of polymerization (DP) is the ratio between the molecular weight of the cellulose molecule and that of a single glucose unit, i.e. DP is the number of units in chain. The DP of native cellulose can vary roughly between 9000 and 15,000, depending on plant material, and the DP of non-degraded modern paper is usually 1200–1500. Typically one of the main effects of the natural aging of cellulose-based artifacts, is the further decrease of DP caused both by acidity and oxidation, and low DP values correspond macroscopically to the worsening of the mechanical properties of cellulose, which can even result in the rupture of fibers and crumbling of sheets. When assessing the degradation status of cellulose, it is practically often preferred to determine the percentage of glycosidic bonds scissored by degradation processes (% S), in order to normalize the DP results obtained from different techniques.

The hierarchical structure of cellulose, starting from polymer chains, is schematically shown in Fig. 5.1b. The supramolecular structure of the polymer is characterized by the presence of a large number of hydrogen bonds, either formed between different chains (intermolecular bonds) or between hydroxyl groups of the same chain (intramolecular bonds). The –OH groups can also form hydrogen bonds with water molecules, in fact cellulose is highly hygroscopic and absorbs or emits water depending on oscillations in relative humidity (RH) and temperature, the average water content being 6–7 % of the polymer weight at standard conditions. The role of hydrophobic interactions and the amphiphilic character of cellulose are still being investigated, and recently the literature reported a debate on these topics as related to the polymer's solubility or insolubility characteristics (Medronho et al. 2012; Glasser et al. 2012).

Cellulose is semi-crystalline (crystallinity degree of 50–90 %): highly crystalline regions exhibit more compact structures and higher resistance to degradation, while amorphous regions are more easily attacked through chemical, physical and biological processes due to acids, oxidative compounds, light, temperature, relative humidity and microorganisms.

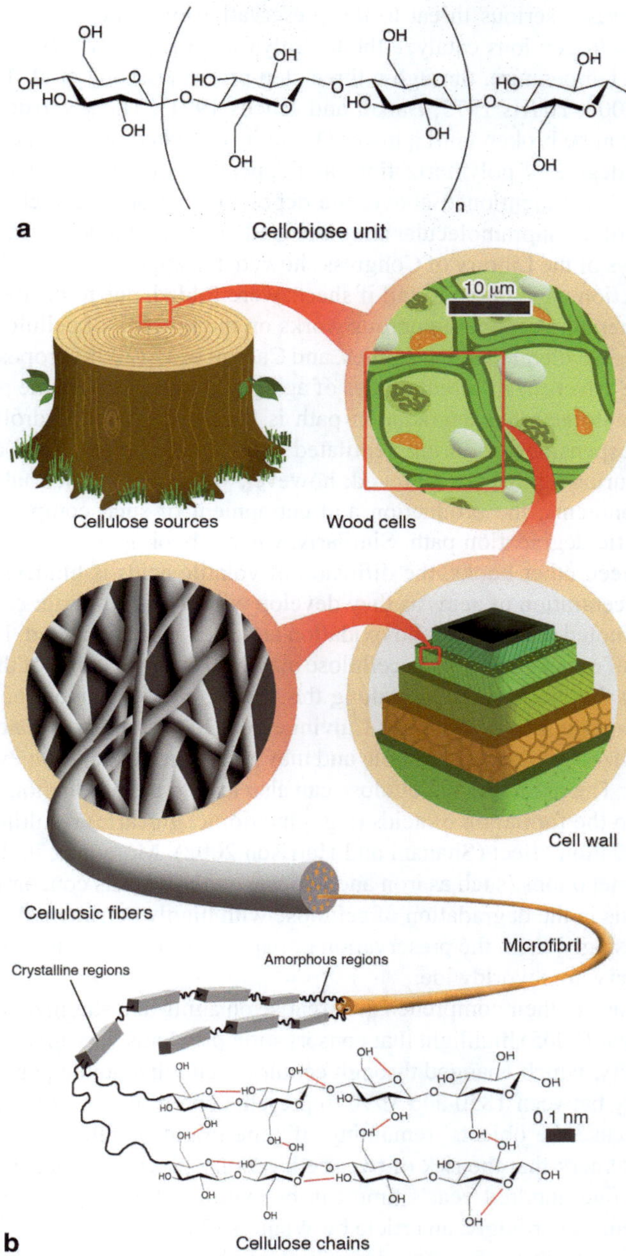

Fig. 5.1 **a** Cellobiose, the formally repeating unit of cellulose chains. **b** Schematic representation of the hierarchical structure of cellulose, from the macro- to the molecular scale. The intermolecular and intramolecular hydrogen bonds are represented by red dashes. (Image by Michele Baglioni)

Acidity poses a serious threat to the preservation of cellulose-based works of art. In fact, hydrogen ions catalyze the hydrolysis of the β-(1,4)-glycosidic bonds, even at room temperature, through a three-step process (Zhang et al. 1994; Lundgaard et al. 2004; Harris 1975; Banait and Jencks 1991). The glycosidic bonds of the polymer can be broken until a lower DP limit is reached, which is called LODP (leveling-off degree of polymerization) and depends on the cellulose composition (Calvini 2005). As mentioned above, the depolymerization of the chains and the modification of the supramolecular structure lead to a loss of mechanical resistance. In fact, surveys of the Library of Congress showed that approximately 25 % of items in their collection were likely to fail if sheets were folded, due to brittleness (Anon 1988). The literature reports numerous works on the kinetics of cellulose degradation, including accelerated aging studies, and Calvini et al. (2008) proposed a model to explain the effects of different modes of aging that greatly affect the process. For instance, an autoretardant degradation path is followed if acid hydrolysis occurs in aqueous suspensions or during ventilated aging where volatile acid compounds (developed during aging) are removed; however, when degradation takes place in sealed environments, the production and entrapment of such compounds lead to an autocatalytic degradation path. Similarly, when a book is on a shelf and tightly inserted between other books, the diffusion of volatile acids is hindered, resulting in higher concentration of acids as they develop within paper (Carter et al. 2000).

It must be noted that acidity and oxidation are often interconnected. For instance the presence of oxidized groups in cellulose may lead to the opening of the rings and to changes in the electron density along the chains, which weaken the glycosidic bonds and favor their hydrolysis. As Calvini et al. (2008) reports, "in acidic conditions the oxidized groups are unstable and may cause further scission even at room temperature". The oxidation of cellulose can also lead to the production of carboxyl groups, and to the formation of acids (e.g. glucuronic, glucaric) resulting in a "spiraling" degradation effect (Shanani and Harrison 2002). Moreover, in the presence of transition metal ions (such as iron and copper) oxidation acts concomitantly with acid hydrolysis in the degradation of cellulose with highly detrimental effects.

Not surprisingly, then, the preservation of paper objects represents a central concern for conservators worldwide.

In the preface to their comprehensive treatise on aging and stabilization of paper, Strlic and Kolar (2005) highlight that conservation problems vary greatly according to paper quality, which changed through centuries, and that acidic paper, produced approximately between 1850 and 1990, represents one of the most pressing issues since in this case the objects' remaining lifetime might be reduced to 100 years. The authors remark that the risk of twentieth century potentially becoming "a new dark age" in five hundred year's time can be avoided if appropriate countermeasures are taken. Accordingly, an article by Wouters (2008) on Science reported that, based on paper condition surveys, about 30 % of objects in Western libraries are in poor condition and in need of intervention, and another 30 % is to reach the same condition by the end of this century. Besides, authors from different countries such as for instance Russia, Iran and China, have reported on the severity of this threat (Dobrodskaya et al. 2004; Afsharpour and Hadadi 2013; Yamjuan et al. 2013). The development of effective materials to counteract acidity and oxidation appears thus as crucial.

One of the main sources of acidity is the presence of aluminum sulfate (added during the papermaking process). Moreover, from the mid-nineteenth century until the latest twentieth century the process included the use of alum in the sizing step, since it promotes the precipitation of the hydrophobic rosin size within the fibers. Alum is hydrolyzed by water, and the hexaaqua–aluminum complex is acidic (K_a similar to that of acetic acid) as reported below:

$$\left[Al(H_2O)_6\right]^{3+} + H_2O \rightleftarrows \left[Al(OH)(H_2O)_5\right]^{2+} + H_3O^+$$

Acidity is also due to residual chemicals from pulping and bleaching, to oxidation products of the lignin component, and to migration of acids from storage materials (cardboard) or printing media. Furthermore, air pollutants such as sulfur dioxide (SO_2) and nitrogen oxides (NO_x) are absorbed by paper and react with moisture within the fibers to form acids in situ (Carter 1989).

Metal gall inks (widely used for centuries) degrade cellulose both through hydrolysis and oxidation. During the making of inks, gallic acid (formed from tannins extracted from gall-nuts) reacts with iron(II) sulfate to produce sulfuric acid and a complex of iron(III). Iron ions promote then the formation of oxidizing radicals and hydrogen peroxide (Neevel 1999):

$$Fe^{2+} + O_2 + H^+ \rightarrow Fe^{3+} + HOO\cdot$$

$$Fe^{2+} + HOO\cdot + H^+ \rightarrow Fe^{3+} + H_2O_2$$

$$Fe^{2+} + H_2O_2 \rightarrow Fe^{3+} + OH^- + HO\cdot \quad \text{(Fenton reaction)}$$

It must be noted that paper contains components that can reduce the oxidized transition metal ions (e.g. from iron(III) to iron(II) ions), initiating a cyclic process.

Copper salts, often contained as impurities in iron-gall ink, are involved in similar degradation mechanisms:

$$Cu^+ + H_2O_2 \rightarrow Cu^{2+} + OH^- + HO\cdot \quad \text{(Fenton-like reaction)}$$

In fact, it has been shown that copper ions are significantly more active than iron ions (Šelih et al. 2007).

As a result, metal gall inks can lead to severe corrosion of paper (up to perforation of the inked areas) and loss of elasticity and tensile strength, as typically observed in documents and manuscripts, e.g. from the sixteenth to eighteenth century.

Conservation science has been developing numerous solutions to address all the aforementioned degradation processes, and to provide stabilization of paper. Indeed, as long as preventive measures can be used to provide appropriate storage conditions (for instance using air purification to reduce gas pollutants), the only way to grant preservation of the threatened patrimony is to apply materials that are specifically tailored to counteract acidity and oxidation. An exhaustive review of the state-of-the-art measures against degradation of paper-based materials has been provided by Banty et al. (2010). In this section, we provide a brief overview of the main methods reported by the literature, ending with the rationale for the use of non-aqueous dispersions of alkaline nanoparticles, which is the focus of this Chapter.

Aqueous solutions of bicarbonates (pH $\approx$ 8.5) or hydroxides (e.g. $Ca(OH)_2$, pH $\approx$ 10.5) have been used for decades, and are still used, for the deacidification of paper, for instance by immersion or spraying (Barrow and Sproull 1959). Besides the direct washing of acidic components and hydro-soluble dirt, the use of aqueous alkaline solutions involves the presence of free, highly mobile hydroxyl ions that allow fast neutralization of acids. However, free OH^- ions, when in excess, are immediately available for interacting with cellulose chains, which can result in depolymerization of aged, oxidized cellulose even at room temperature (Calvini et al. 1988). One of the main processes is the β–alkoxy elimination that produces the peeling of cellulose chains (Santucci and Plossi Zappalà 2001). Moreover, the aqueous alkaline environment can alter inks, e.g. through leaching or discoloration. Water also causes the swelling of the cellulose fibers, and the migration of corrosive metal ions during the treatments (Henniges et al. 2006).

C. Williams and G.B. Kelly invented a gaseous treatment (based on diethyl zinc) in the 1970s, and developed the process through the early 1980s. However the utilization of this technology for mass deacidification involves high costs and significant safety and maintainance precautions (diethyl zinc reacts explosively with oxygen), which hinder the commercialization of the process (Harris and Shahani 1994).

Several non-aqueous wet deacidification treatments have been thus developed in the last 50 years. The Wei t'O method, developed by R. Smith in the 1970s, consists in the application of a solution of magnesium methoxy methyl carbonate in a mixture of methyl alcohol and freons, later replaced by hydrochloro-fluoro carbons. Following the application by spraying or immersion, the methoxy methyl carbonate is hydrolyzed by moisture to form magnesium hydroxide, methyl alcohol and CO_2 (Burgess et al. 1992). Other methods have been developed, still focusing on the hydrolysis of Mg-organic salts to obtain the in situ formation of $Mg(OH)_2$, such as the Sable, the CSC Book Saver and the Papersave, which was developed by Battelle Ingenieurtechnik GmbH (hence the name "Battelle process") and uses a complex of magnesium and titanium alcoholate in hexadimethyldisiloxane (HMDO). The magnesium alcoholate neutralizes the free acids in paper, and during reconditioning the excess is converted into $Mg(OH)_2$ that serves as a buffer against recurring acidity. The titanium alcoholate must be included in the applied solution because magnesium alcoholate is insoluble in HMDO. Upon reaction with moisture, titanium alcoholate turns into titanium hydroxide, $Ti(OH)_4$, which then decomposes into TiO_2 and water, therefore the titanium alcoholate does not eventually contribute to deacidification (Blüher and Grossenbacher 2006).

One of the most widely used non-aqueous methods is the Bookkeeper (Preservation Technologies L.P.—PTLP), which uses microparticles of MgO dispersed in a fluorinated solvent. Different products have been developed for use in closed mass deacidification machines or for open spray application. The spray reagent basically consists of MgO particles (4.3 g/L), a blend of C_5–C_{18} perfluoroalkanes (with < 1 % of byproducts) and several additives (Zumbül and Wuelfert 2001). The latter include a surfactant (< 0.1 %), probably either a perfluorinated Mg-soap, a perfluoropolyether derivative or a non-ionic fluorinated acyl ester; in addition, nonafluoromethoxybuthane is also present. Following application, the MgO particles hydrolyze to form $Mg(OH)_2$ which provides deacidification, and any excess remains as an alkaline reserve, eventually transforming into carbonate. Bookkeeper has proven to be a very good method, owing to its effectiveness and several advantages such as the ease of application, no required preconditioning, solvent inertness (non-swelling, chemical inertness) and a controlled hydrolysis process that grants neutralization without exposing cellulose to strong alkalinity. Zumbül and Wuelfert (2001) clarified that the additives are adsorbed on the solid particles, and due to their hydrophobicity they act as retarders for the hydrolysis of MgO. The slow formation of magnesium hydroxide and, consequently, magnesium carbonate, might be risky for sensitive paper substrates that might be damaged by the high alkaline pH developed by MgO and $Mg(OH)_2$ particles (a too highly alkaline environment might degrade cellulose fibers due to β–alkoxy elimination as mentioned above). Although the Bokkeeper is one of the more reliable methodologies for the deacidification, a limitation in the general application of this method is that the fluorinated materials used in large amount as additives to stabilize the particles remain included in the alkaline reserve of hydrated magnesium carbonates, and also deposit as residues on the treated surface. Finally, elsewhere it has been reported that the application on low porosity paper can lead to the formation of veils, due to the poor penetration of the microparticles within cellulose fibers (Stauderman et al. 1996; Pauk 1996), and the method is normally not recommended on slick, highly calendered paper, and on black or dark colored substrates (Boone et al. 1998).

Some researchers in the last 15 years also proposed the use of supercritical carbon dioxide (CO_2SCF) as solvent system for alkaline or strengthening agents, based on the non-toxicity, non-flammability and relatively low costs of this approach. Materials such as $CaCO_3$, magnesium alkoxides, methoxycarbonate, catechol and borax have been tested with promising results (Selli et al. 2000; Dobrodskaya et al. 2004; Yamjuan et al. 2013).

For what concerns anti-oxidants, one of the most effective aqueous methods involves treatment of paper with a solution of myo-inositol hexaphosphate (phytate) (Neevel and Reissland 1997, Neevel 1999). Phytate is a natural anti-oxidant that complexes iron(II) ions and blocks the production of hydroxyl radicals from H_2O_2 (Fenton reaction). The treatment is followed by deacidification with an aqueous solution of calcium bicarbonate. Kolar et al. (2007) proposed a further development of the method, through the use of a magnesium phytate solution, which grants several practical advantages. Besides the use of an aqueous environment, another possible limitation is that complexing agents such as phytate are usually metal-specific. However it has been reported that inositol hexaphosphates have pronounced binding capacity for copper(II) at pH 5–7 (Persson et al. 1998), which is useful to inhibit the reduction to copper(I) that re-initiates Fenton-like reaction. In fact, Henniges et al. (2008) showed that treatment with phytate was beneficial against degradation caused by inks containing both iron and copper ions.

Alternatively, Malesic et al. (2005) proposed the use of halides as radical scavengers that react with hydroxyl radicals in a large T and pH interval. The authors showed that the effectiveness of stabilization depends on the size of the cation, and the best result was achieved using tetrabutylammonium bromide. A promising update was provided by Kolar et al. (2008), who investigated the use of alkylimidazolium bromides, in combination with alkali such as magnesium ethoxide in ethanol, which proved better stabilization of iron gall inked model papers than both the previously studied bromides and phytates. An alcohol solution of tetrabutylammonium bromide and magnesium ethoxide gave the best results on models containing corrosive copper pigments. The authors concluded that the effects of the treatment solutions must be studied also on historical samples before application on historical artifacts. According to Banty et al. (2010), calcium and magnesium alkoxides in conjunction with quaternary ammonium bromides are being developed for the stabilization of iron gall inked paper.

As demonstrated by the literature cited, research work in the field of paper preservation is constantly dedicated to finding methods for addressing all the conservation issues simultaneously, in one step. The ideal treatment must avoid the drawbacks discussed in the previous paragraphs, namely the leaching or removal of original parts of the artifacts (e.g. inks or paints), the formation of veils or aesthetic alterations on the surface of the treated objects, and the use of toxic and environmentally unfriendly solvents. Moreover, the treatment should involve low costs and be practically feasible.

Based on these requirements, starting since the early 2000s researchers at CSGI (Consorzio Interuniversitario per lo Sviluppo dei Sistemi a Grande Interfase; Research Center for Colloid and Surface Science) explored the potential of nanoparticles of alkaline earth metal hydroxides for the deacidification and the stabilization of paper objects, and the method has been patented in 2002 (Baglioni et al. 2002).

In fact, calcium hydroxide is an excellent deacidifying agent but the use of aqueous solutions of $Ca(OH)_2$ is too chemically aggressive on paper fibers, as discussed above. A valid alternative is to deliver the hydroxide in the form of solid particles, dispersed in an appropriate solvent, onto the cellulose fibers. This allows a gradual and safe release of OH^- ions from the particles, rather than the sudden delivery of free, highly mobile ions as occurs in the aqueous solutions. The neutralization between the OH^- ions (from particles) and the H^+ ions (coming from acidic species or groups) takes place on the surface of the particles through a layer of water that comes from environmental humidity and fibers moisture, as described by the following basic reaction:

$$H^+(aq) + OH^-(aq) \rightarrow H_2O \tag{5.1}$$

It has been hypothesized that calcium or magnesium ions (from the deacidification treatment) might interact with carboxylate groups of cellulose (see Fig. 5.2) (Baglioni et al. 2012). Elsewhere, Bukovský (2000) hypothesized the interaction of cel-

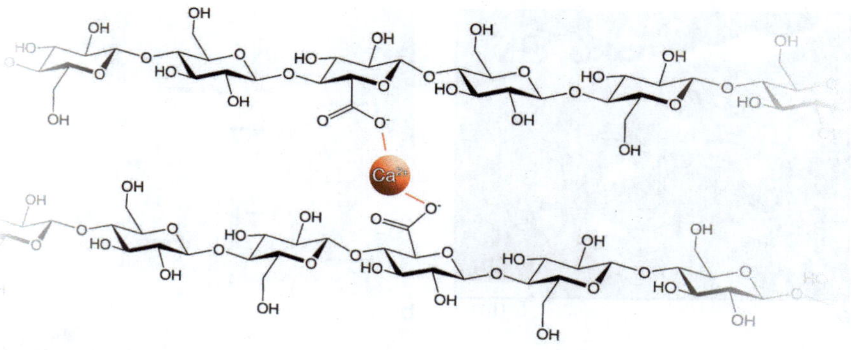

Fig. 5.2 Possible interactions between Ca^{2+} ions (coming from $Ca(OH)_2$ nanoparticles) and carboxylate groups ($-COO^-$) on the cellulose chains, following the neutralization treatment. (Image by Michele Baglioni)

lulose hydroxyl groups (R-OH) with magnesium methoxy methyl carbonate, giving for instance R_1–O–Mg–O–R_2 bonds, to explain the increased folding endurance of paper. Similarly, an enhancement of the resistance to folding has been also observed on paper samples treated with dispersions of calcium hydroxide nanoparticles.

The $Ca(OH)_2$ particles surface that is not consumed during the neutralization process reacts with CO_2, turning into $CaCO_3$, as described in Chap. 2 (Sect. 2.3). Therefore, the carbonated particles remain as a buffer against recurring acidity—in fact paper that contains calcium carbonate fillers exhibits better storage characteristics (Hubbe 2005). Moreover, hydroxide carbonation is another important factor, together with the reduced mobility of OH^- ions, in preventing the formation of an excessively alkaline environment that could be harmful to cellulose.

Stable dispersions of calcium or magnesium hydroxide nanoparticles in short chain alcohols have been prepared in the last decades using different synthetic processes, as discussed in Chap. 2. The application of dispersions on paper, either through brushing, spraying, dripping or immersion, allows the simple delivery of the desired amount of hydroxide without using any precursors such as magnesium oxide, alkoxides or methoxy methyl carbonate. Moreover, the nanoparticles are stably dispersed in short chain alcohols without using any additives or surfactants, so to further minimize unforeseen effects. It has been shown that upon evaporation of the solvent, the nanoparticles adhere to the cellulose fibers (see Fig. 5.3), which is explained considering that the positive charge on the surface of the particles promotes their interaction with the hydroxyl groups of cellulose (Giorgi et al. 2002a).

As noted by Sequeira et al. (2006), the reduced size of the hydroxide nanoparticles, and consequently their elevated surface area (see Chap. 1, Fig 1.1), involve a high capacity of acid neutralization, and the rapid transformation of the hydroxide excess into carbonate. In fact, the application of $Ca(OH)_2$ nanoparticles to real nineteenth century paper specimens resulted in the neutralization of acidity and in the formation of a $CaCO_3$ buffer, which protected paper against accelerated aging under

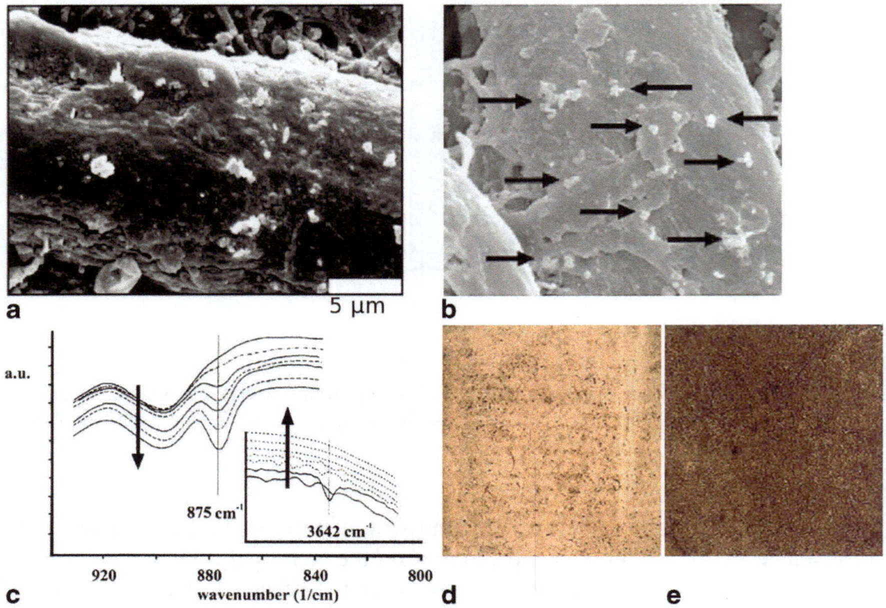

Fig. 5.3 **a, b** SEM image of a nineteenth century paper sample treated with $Ca(OH)_2$ nanoparticles, showing the particles adhered to the cellulose fibers. Bar is 5 µm. **c** The carbonation of nanoparticles as shown by ATR-FTIR spectra of the treated paper sample, collected starting from the application of particles. Arrows indicate the direction of the time evolution, showing the decrease of the $Ca(OH)_2$ absorption band at 3642 cm^{-1}, and the increase of the $CaCO_3$ band at 875 cm^{-1}. **d** Picture of a nineteenth century paper sample that was treated with the $Ca(OH)_2$ nanoparticles and then aged. **e** Picture of a nineteenth century paper sample that underwent aging without any previous treatment. The dramatic discoloration indicates severe cellulose degradation. (Reprinted with permission from Giorgi et al. 2002a. Copyright © 2002, American Chemical Society)

severe hydrothermal conditions (RH = 85 % and T = 90 °C for 21 days), as shown in Fig. 5.3. Moreover, a study by Poggi et al. (2010) showed that the treatment of metal gall inked paper with a dispersion of $Mg(OH)_2$ nanoparticles in 2-propanol was also highly effective in protecting the fibers of cellulose against degradation. The application of $Mg(OH)_2$ nanoparticles kept the pH stably around 7 and preserved the mechanical properties of the cellulose fibers throughout aging, whereas non-protected samples exhibited strong discoloration and dramatic mechanical failure (see Fig. 5.4). Microscopically, this was due to the fact that the percentage of cellulose glycosidic bonds scissored by degradation processes (% S) was kept low by the treatment with alkaline particles throughout aging, as opposed to the case of inked samples that had not been treated (see Fig. 5.5). The application of the Bookkeeper method, which uses larger particles, on the same samples provided slightly worse results, and it must be noticed that in general the use of nano- (rather than microsized) particles can also be advantageous to achieve better penetration through low-porosity fiber networks, without producing veils due to surface accumulation.

Fig. 5.4 Comparison between metal inked paper samples (Whatman) before and after the aging. (*Left*) A sample not treated with nanoparticles, which did not undergo aging. (*Center*) A sample aged without any stabilization treatment, where the loss of the paper mechanical properties is evident. (*Right*) A sample treated with $Mg(OH)_2$ nanoparticles and aged: the sample retained the original mechanical properties, as before aging. (Reprinted with permission from Poggi et al. 2010. Copyright © 2010, American Chemical Society)

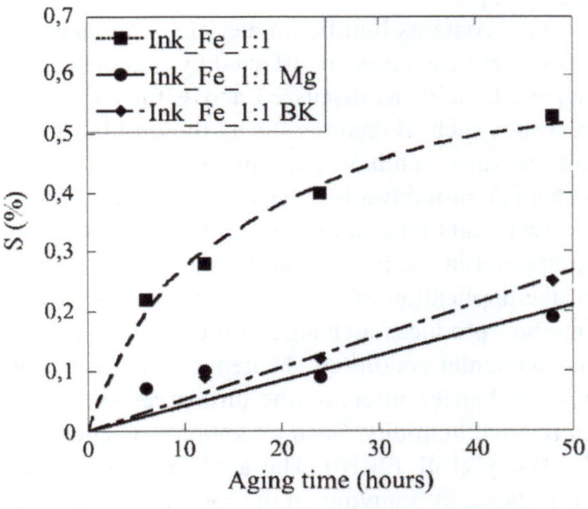

Fig. 5.5 Degradation curves showing the percentage of cellulose glycosidic bonds scissored by oxidation and acid hydrolysis (%S), during accelerated aging of Whatman paper samples. **"Ink_Fe_1:1"** refers to samples treated with a classic iron gall ink. **"Ink_Fe_1:1 Mg"** refers to inked samples that were stabilized with $Mg(OH)_2$ nanoparticles and then aged. **"Ink_Fe_1:1 BK"** refers to inked samples that were stabilized with the Bookkeeper method (spray) and then aged. (Reprinted with permission from Poggi et al. 2010. Copyright © 2010, American Chemical Society)

Giorgi (2013) later showed that the beneficial effect provided by the application of $Mg(OH)_2$ nanoparticles, in terms of reduced % S, was comparable to that obtained by treating inked paper samples with an anti-oxidant treatment based on halides. In fact, Strlic et al. (2003) had previously investigated the redox couples Fe(II)/Fe(III) and Cu(I)/Cu(II) and had showed that the catalytic activity of iron is maximum when pH is 8.0–8.5, while that of copper increases sharply when pH is higher than 8.5. Moreover, the activity for both metals is minimal when pH is around 7.

These results indicated that by stably maintaining the final pH of paper around neutrality, with a single-step deacidification treatment, it is possible to preserve cellulose from the degradation induced both by acidic (e.g. sulfuric acid) and oxidative agents (iron and copper ions). An alkaline buffer can be provided against recurring acidity (e.g. maximum final pH around 8.0–8.5, see Sect. 5.4).

The short chain alcohols used as nanoparticles dispersing media are environmentally friendly solvents with low toxicity, and are inert towards several (ethanol) or numerous (2-propanol) metal gall inks. Other solvents (e.g. hydrocarbons) have been successfully experimented to extend the application to paper containing modern inks.

Overall, the method has thus proved highly feasible due to its high effectiveness, low costs and ease of application. The results obtained on paper paved the way for the application of nanoparticles onto other cellulose-based materials, such as canvas and wood, since the degradation processes that need to be addressed are similar to those experienced by paper.

Linen, for instance, contains lignified materials that make it more prone to deterioration, and in general canvases are affected by photo-oxidation, gas pollutants (SO_2, NO_x) and volatile acids as discussed above for paper. Acidity can also be developed by varnishes such as dammar, or by the oil binder (Brewis and Halsall 1961). As a result, canvases can have pH values as low as 3.5–4.0, and the consequent loss in mechanical properties is enhanced by the tensile strengths the textiles are subjected to. Treatments for cellulose deacidification, originally developed for paper, have been applied in the past to canvas paintings, for instance Foster et al. (1997) evaluated the application of Wei T'o on the backside of canvases and observed that the method produces a change in the stiffness of the samples that depends on the environmental conditions. Moreover, the treatment leaves a coating that acts as a moisture barrier, affecting the further response of the treated canvas to fluctuations in relative humidity. Such alteration was later deemed as a "not intended" effect by Banty et al. (2010). The application of dispersions of alkaline nanoparticles, for instance by spraying on the backside of the painting, is a promising alternative method for the preservation of acidic canvases, given its feasibility and effectiveness as demonstrated on paper objects. In the case of paintings, the main factor to be considered is the compatibility between the particles dispersing medium and the layers of the artifacts (preparation, painted layers), which are typically sensitive to solvents with different polarities.

For what concerns wood, acidity can pose a threat to artifacts in several circumstances. First, it must be noticed that wood itself emits acidic compounds, such as acetic and formic acid. Such emissions promote degradation of objects that are in close contact with the wood surface, for instance artifacts stored or displayed in wooden boxes, and paper artworks or canvases in contact with wooden frames. Degradation also takes place when works of art are exposed to atmosphere containing acidic VOCs (volatile organic compounds) emitted by wooden surfaces. Coatings can be applied on wood to decrease the emission of acids, however it is known that several poly(vinyl acetate) based coatings and adhesives traditionally used for cabinets and cases emit in turn acidic VOCs (Tétreault and Stamatopoulou 1997), and in some cases it is recommended that materials with emission potential of hazardous compounds must be replaced by materials such as metal, glass and powder coatings (Schieweck et al. 2007). Potentially aggressive organic compounds can also be found inside microclimate frames (Lòpez-Aparicio et al. 2010). An elegant alternative solution to this issue is to provide the wooden surface of frames, boxes etc. with an alkaline buffer that neutralizes the acidic emissions directly in situ. In this way, the emission of acidic VOCs is depressed and the use of potentially hazardous coatings is avoided. The treatment of wooden surfaces with a dispersion of alkaline nanoparticles has proven to be a facile method to achieve this task (Giorgi et al. 2009).

Another field where nanoscience has proven effective in counteracting cellulose acidity concerns the preservation of archaeological waterlogged wood. Two famous cases involve the conservation of the wooden ships Vasa and Mary Rose, which are currently being degraded by compounds such as sulfuric acid, oxalic acid and iron salts (Giorgi et al. 2005a; Schofield et al. 2011). According to the literature, one of the origins of acidity is the large quantity of reduced sulfur compounds that accumulated within the timbers while the ships were lying on the seabed in polluted anoxic waters, as a result of the metabolic activity of sulfate-reducing bacteria. Starting from the salvage, the sulfur compounds are being oxidized, producing sulfuric acid and sulfates. Iron ions, coming from the corrosion of nails and other metallic components, promote the oxidation of sulfur and cause the degradation of cellulose, also leading to the formation of oxalic acid from oak wood. However, the complex interaction between the chemical species in wood is still not completely understood, and recent studies on the Vasa have suggested that sulfur may in fact act as an antioxidant, since holocellulose degradation was found to be higher in areas of high iron content but low sulfur content. Moreover, iron(III) ions are probably involved in the degradation of polyethylene glycol (PEG), a consolidant that was extensively applied to prevent wood shrinkage after the salvage. The degradation of PEG is still debated: some researchers claim that the deterioration is negligible while others highlighted that the process might take place through acid hydrolysis. The product's degradation might eventually contribute to the formation of formic acid.

Overall, these interconnected degradation processes lead to very low local pH values (e.g. 0–4), whereas the pH of fresh oak wood is typically around 4.5, and that of pine is ca. 5.5. Acid hydrolysis of cellulose might cause structural failures that are potentially disastrous on such large objects. Even though immediate danger is not foreseen, it is important to understand and counteract wood degradation to preserve the ships for future generations. Therefore, several research groups are involved in finding possible solutions against the action of acids and iron ions. The main issue common to all methods developed so far is that access to the inner layers of the PEG-impregnated wood is problematic. Besides, conventional alkaline aqueous solutions involve similar issues as those discussed for paper deacidification. Alternatively, ammonia gas treatment has been tested on Vasa samples, but limitations regard the maximum penetration achieved (0.5–1.0 cm for oak samples, up to 3.5 cm for pine) and extensive cellulose degradation for Vasa samples without PEG, while PEG-treated samples were not degraded (Fors and Richards 2010). A treatment based on aqueous solutions of chelating agents such as DTPA (diethylenetriamine pentaacetate) has been developed to extract iron, sulfur and old PEG from wood. However, the pH required for the full efficiency of the method (9–11) is too risky for wood, and more neutral pH values slightly decrease the extraction effectiveness. Moreover, such treatments might require several years and the iron extraction might not access the inner regions of the timber (< 1–2 cm depth).

The use of alkaline earth hydroxide or carbonate nanoparticles has been proposed to stably neutralize pH (inhibiting iron ions catalytic activity) and prevent the formation of sulfuric acid. For instance, treatment with $Mg(OH)_2$ nanoparticles resulted in increasing the pH of acidic wood samples up to depths of 2 cm, for at least 1 year (Chelazzi et al. 2006). Application of nanoparticles under vacuum or moderate pressurization on samples, using $Ca(OH)_2$ particles of ca. 80 nm, resulted in increased penetration up to ~ 10 cm.

As a general remark, the treatment of smaller wooden objects appears as more practical regardless of the method used. Obviously, large-scale treatments (including spraying or immersion) of huge and complex objects such as a ship's hull would involve technical and monetary issues, and thus remain as open issues.

Finally, it is worth noticing that the use of nanoparticles has been experimented also for other tasks (besides pH control) in the field of conservation of cellulose-based works of art. Some of the proposed applications include: (i) the use of TiO_2 or ZnO nanoparticles for the protection of paper supports against soiling, microorganisms and UV (Afsharpour and Hadadi 2013; El-Feky et al. 2014); (ii) the application of silica and other inorganic nanosols for the treatment of wood, so to improve mechanical properties, or as flame retardants, antimicrobial coatings, etc. (Mahltig et al. 2008).

The following sections will deal with the different formulations of particles dispersions that have been produced and used for deacidification purposes, and on the possible application fields. Finally, practical guidelines will be provided, based on the experience acquired in the last decades.

5.2 Formulations

The main $Ca(OH)_2$ nanoparticles dispersions for application to works of art have been discussed in detail in Chap. 2 (Sect. 2.2). This section focuses on the formulations that have been developed and used in the last decades for pH control on cellulose-based substrates.

The first application of calcium hydroxide nanoparticles for paper preservation dates back to the early 2000s (Giorgi et al. 2002a, b). Nanoparticles prepared through homogeneous phase bottom–up processes and dispersed in 2-propanol were used to deacidify paper dating from the fourteenth, seventeenth, nineteenth, twentieth century. The fourteenth–nineteenth century samples were rag paper, with

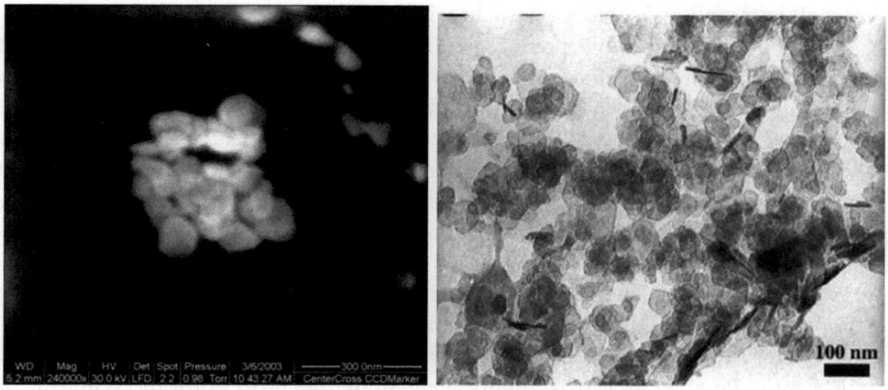

Fig. 5.6 $Mg(OH)_2$ nanoparticles obtained through a homogeneous phase bottom–up process, starting from magnesium sulfate and sodium hydroxide. (*Left*) SEM image, bar is 300 nm. (*Right*) TEM image, bar is 100 nm. (Reprinted with permission from Giorgi et al. 2005b. Copyright © 2005, American Chemical Society)

sizing of either starch or gelatin. The twentieth century sample was wood pulp paper, with rosin sizing. The nanoparticles granted effective deacidification for long period of time under natural aging, and a significant resistance to accelerated aging under severe conditions (T = 90 °C, RH = 85 %, for 21 days), as shown in Fig. 5.3.

Few years later, Giorgi et al. (2005b) prepared and applied $Mg(OH)_2$ nanoparticles dispersions in 2-propanol for paper deacidification. The rationale was that several well established deacidification methods, such as the Wei T'o and the Bookkeeper, used precursors for the formation of magnesium hydroxide (and then carbonate) within the paper fibers, therefore the direct application of $Mg(OH)_2$ nanoparticles (without any precursor or additives) represented a much simpler option. Besides, the reduced particle size was expected to result in higher reactivity, in terms of neutralization of acids and carbonation of hydroxide, as explained in the previous section. The particles were prepared through a homogeneous phase process starting from magnesium salts with several different counterions, and it was found that the final particle size increases from ca. 50 to 200 nm following the Hofmeister anion series: sulfate<chloride<nitrate≤perchlorate. For instance, particles of ca. 40–100 nm were obtained starting from magnesium sulfate and sodium hydroxide (see Fig. 5.6). Paper samples treated with $Mg(OH)_2$ nanoparticles (simply by brushing) were compared to those treated with $Ca(OH)_2$ nanoparticles or with the Wei T'o method, measuring the percentage of broken bonds and the tensile strength resistance of paper after photo-oxidative and hydrothermal aging. In fact, results indicated that magnesium and calcium hydroxide nanoparticles granted similar protection to paper, and both were more effective than the Wei T'o.

In the following years, the beneficial effects of treatment with dispersions of either $Ca(OH)_2$ or $Mg(OH)_2$ nanoparticles have been further assessed, including for instance protection of aged samples throughout aging cycles, compatibility with iron gall inks, and satisfactory photochemical stability of the treated paper (Sequeira et al. 2006; Stefanis and Panayiotou 2007, 2008, 2010). Poggi et al. (2010, 2011) highlighted the importance of tuning the final paper pH around 6.5–7.5 in order to achieve the concomitant neutralization of acidity and inhibition of the oxidative activity of the iron and copper ions in metal gall inked paper, as discussed above. The dispersions of calcium hydroxide nanoparticles were obtained either through a bottom–up and a top–down process (starting from CaO, as described in Chap. 2, Sect. 2.2.1) and magnesium hydroxide nanoparticles were prepared through a bottom–up process.

As mentioned in Chap. 2, a commercial formulation of calcium hydroxide nanoparticles in 2-propanol is available on the market since 2008 under the trademark Nanorestore® (from: CTS Srl., Altavilla Vicentina/Italy. www.ctseurope.com).

Recently, Poggi et al. (2014) reported an alternative process for the synthesis of stable and high concentrated dispersions of $Ca(OH)_2$ nanoparticles (ca. 35 g/L) (see Chap. 2, Sect. 2.1.1). The particles are dispersed in ethanol or 1-propanol, the two systems exhibiting different size distributions (see Fig. 5.7). These formulations have been tested on acidic paper and canvas samples (Poggi et al. 2014), and results indicated that the particles are highly effective in stabilizing the pH of both paper and canvas around neutrality even throughout aging under severe hydrothermal conditions (e.g. 4 months at $T = 80\,°C$ and $RH = 75\,\%$), and cellulose depolymerization is consequently inhibited. A commercial formulation of calcium hydroxide nanoparticles for the preservation of paper has been registered in 2014 under the trademark Nanorestore Paper®, and is available on request from CSGI.

Finally, it is worth mentioning that different solvents, besides short chain alcohols, have been considered for the treatment of documents containing sensitive inks such as ballpoint pen inks and others. For instance, dispersions of $Ca(OH)_2$ nanoparticles in cyclohexane have been developed as possible alternatives. Such dispersions exhibit lower kinetic stability than systems in alcohols, but allow the effective treatment of documents that contain modern inks.

Calcium hydroxide nanoparticles in 2-propanol were highly effective in decreasing the emission of acetic acid from oak wood, as anticipated in the previous section. Figure 5.8 shows that acid emission was depressed ($<20\ \mu g/m^2\ h$) as compared to untreated wood (Giorgi et al. 2009). In this case, fuming with ammonia did not show any beneficial effect, since emissions of acetic acid remained as high as from the untreated oak sample. These results open for the potential application of alkaline nanoparticles on frames, display boxes etc., and in general for preventing damage to any object in close contact or near proximity to the treated wood surface. For instance, it has been shown that acetic and formic acid vapors emitted from wooden parts in historical church organs are corrosive for the lead pipes. In fact, the concentration of acid VOCs was found to be higher inside the organs than in the church environment. The application of nanoparticles represents a valid and more wood-compatible alternative to the use of synthetic coatings on the wooden surface.

For what concerns application on archaeological and waterlogged wood, both dispersions of calcium and magnesium hydroxide nanoparticles in 2-propanol have

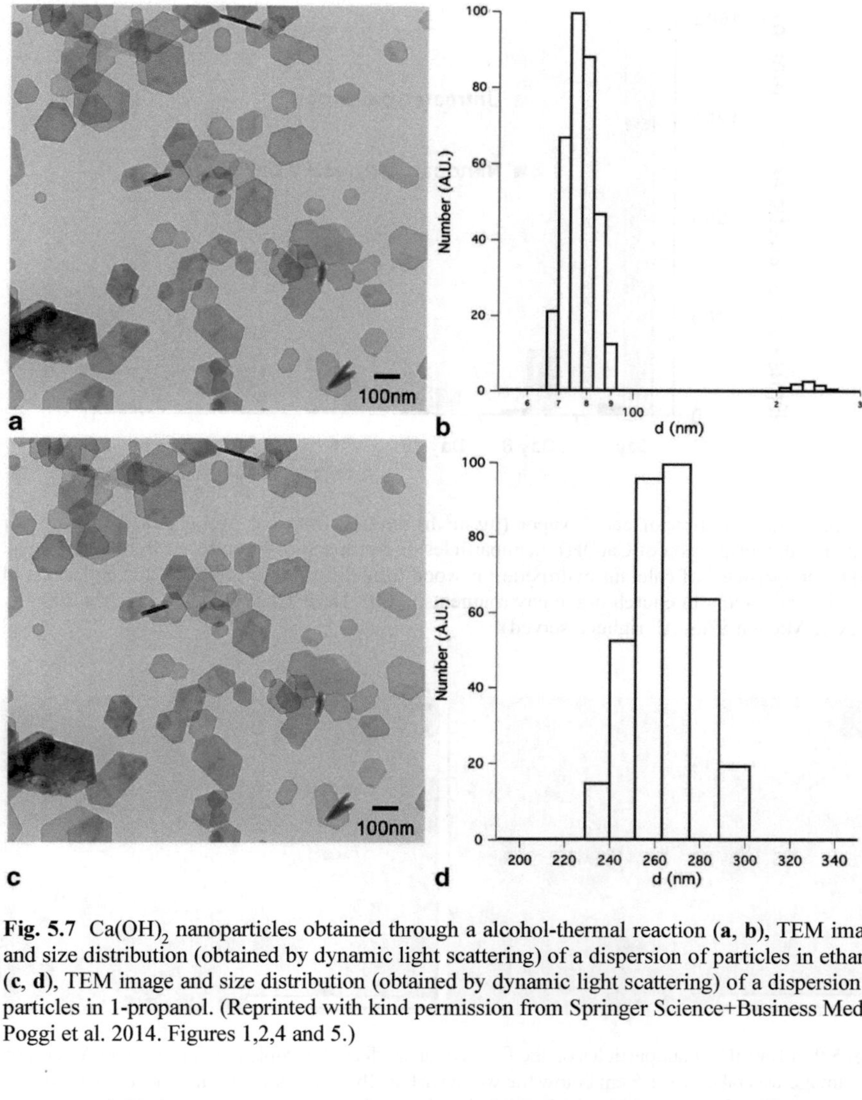

Fig. 5.7 $Ca(OH)_2$ nanoparticles obtained through a alcohol-thermal reaction (**a**, **b**), TEM image and size distribution (obtained by dynamic light scattering) of a dispersion of particles in ethanol (**c**, **d**), TEM image and size distribution (obtained by dynamic light scattering) of a dispersion of particles in 1-propanol. (Reprinted with kind permission from Springer Science+Business Media. Poggi et al. 2014. Figures 1,2,4 and 5.)

been tested on acidic samples, including heavily degraded oak and pine samples coming from the Vasa ship (Chelazzi et al. 2006; Giorgi et al. 2005a, 2006). As discussed in the previous section, the main issue in this case is to achieve sufficient penetration through the PEG-impregnated wooden matrix, which is a common limitation to all the methods that have been developed so far. In fact, the application of nanoparticles can be considered as a promising method in that it is more effective and less aggressive than aqueous alkaline treatments such as washing soda/baking soda solution poultices. It has been shown that hydroxide nanoparticles are effective in adhering to wood fibers, counteracting acidity also on the long-term (during both

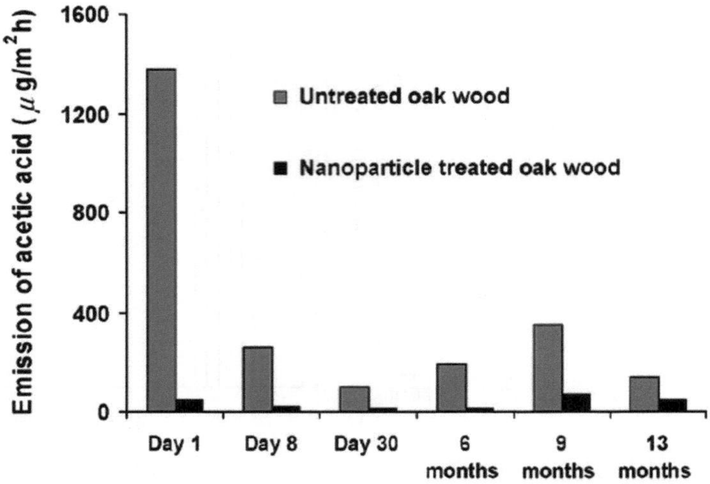

Fig. 5.8 Concentration of acetic vapor (μg/m^2 h) emitted from oak wood either untreated or treated with a dispersion of $Ca(OH)_2$ nanoparticles in 2-propanol. (Reproduced from Giorgi et al. 2009 Nanoparticles of calcium hydroxide for wood deacidification: decreasing the emissions of organic acid vapors in church organ environments. J Cult Herit 10:206–213, Copyright 2009 by Elsevier Masson SAS. All rights reserved)

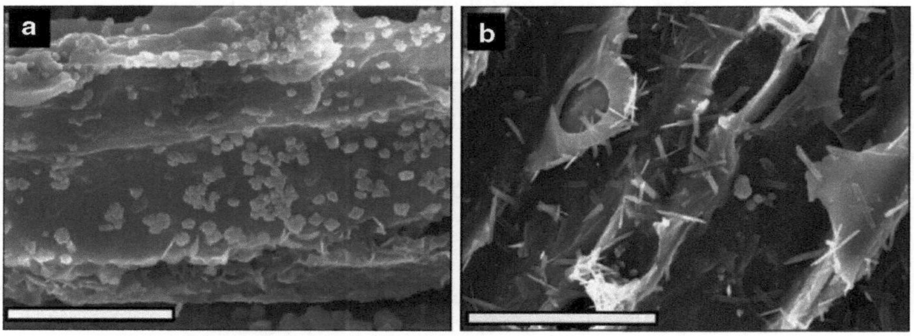

Fig. 5.9 **a** $Ca(OH)_2$ nanoparticles on the fibers of an acidic oak sample coming from the Vasa ship. The image was taken ca. 1.5 cm below the wood surface (bar = 20 μm). **b** Calcium sulphate formed after deacidification (bar = 20 μm). (Reprinted with permission from Giorgi et al. 2005. Copyright © 2005, American Chemical Society)

accelerated and natural aging), and the sulfate crystals that form upon neutralization of sulfuric acid are small enough to avoid mechanical stress in the wood pores (see Fig. 5.9). The hydroxide particles excess will eventually undergo carbonation; moreover dispersions of calcium carbonate nanoparticles can also be used.

Another promising formulation, proposed for the conservation of the Mary Rose timbers, use strontium carbonate ($SrCO_3$) particles with crystallites of 20–50 nm, obtained with a top–down process (high energy ball milling) starting from commercial $SrCO_3$ (Schofield et al. 2011; Chadwick et al. 2012; 2014). The particles are dispersed in 2-propanol through sonication. Wood samples were soaked in the dispersions and further sonicated for 3 days. Besides some deacidification effect, the main rationale for the use of $SrCO_3$ particles is their influence on the reduced sulfur species and iron compounds in the wood. Results of spectroscopic analyses showed that the treatment caused a decrease of the sulfur reduced species in wood and the formation of both sulfoxide and strontium sulfate ($SrSO_4$), which is stable and insoluble. Moreover, pyrite slowly converts into iron carbonate. Consequently, the further formation of sulfuric acid in the wood matrix is hampered. The main limitation is that using the ball-milling process, although crystallites from 20–50 nm are present, the particles size is in the micro-size range as confirmed by the low penetration of particles (2 mm using dispersions in 2-propanol) and low diffusion coefficients through fresh oak wood (order of magnitude of 10^{-13} m^2s^{-1} for 2-propanol dispersions). The presence of PEG slows the process. An additional side effect in the use of strontium particles is that, differently from calcium and magnesium, strontium ions exhibit moderate toxicity.

5.3 Application Field

In general, the application of calcium and magnesium hydroxide (or carbonate) nanoparticles is indicated as an effective tool to counteract the degradation of cellulose-based artifacts, also in the presence of metal ions such as copper and iron, which are involved in catalytic oxidative processes. The preservation of paper, canvas and wood are all practical application fields, but the method can be potentially extended to different substrates.

When evaluating the applicability of nanoparticles dispersions to inked paper, one of the main factors to be considered is the compatibility of the dispersing medium with the inks present on the manuscripts. As previously anticipated, 2-propanol generally exhibits good compatibility towards metal gall inks. However, inks recipes vary from place to place and at different times in history. Therefore, it is advisable to check the possible occurrence of ink bleeding or alteration by tests on small spots, before applying the dispersion. Similar considerations and precautions apply to ethanol that, being more polar than propanol, poses some higher risks to metal gall inks. Nonetheless, ethanol dispersions have been safely used on several occasions.

The application of propanol and ethanol dispersions is less straightforward on modern inks, which are sensitive to such solvents. However, it must be considered that the sensitiveness of inks depends not only on the chemical nature of the solvents, but also on the application method. For instance brushing combines the solving power of solvents with mechanical action, and thus enhances risks of ink bleeding. Recently, conservators at the historical archives of the National Museum of Anthropology in Mexico City carried out extensive tests to assess the effects of different methods for applying 2-propanol dispersions of $Ca(OH)_2$ nanoparticles on documents containing modern inks (e.g. ballpoint pen ink, printing ink, stamps). Tools such as sprayer, nebulizer and suction table were used as compared to the brushing method. Results indicated that direct application of the dispersion on inked paper using a nebulizer was safe in 66 % of cases, and maintained an effective neutralization of acidity. As noted by the conservators, this method could be applied on extensive areas using semi-industrial machines that allow working at a constant distance from the document (e.g. 5 cm). The nebulizer can also be used to saturate the environment in a close box containing the document, which avoids leaching of inks, but also results in a slower neutralization of acids. The spraying method is also effective in decreasing the risks of solvent alteration. Finally, as mentioned in the previous section, hydrocarbons such as cyclohexane can be used as alternative dispersing media, allowing safe treatment even through brushing in most cases.

The applicability of dispersions to canvas paintings is wide and not yet completely explored. So far, several tests and even full treatments by spraying have been carried out on the backside of classic oil paintings (e.g. nineteenth century) with positive results, using dispersions of calcium hydroxide nanoparticles in propanol, and no detrimental effects have been observed. The applications lead to neutralization of acidity and seem to provide some increase in the mechanical resistance of canvas. As a general criterion, 1-propanol is to be preferred to ethanol in this case, owing to the sensitivity of both the paint and ground layers to polar solvents (Poggi et al. 2014).

Despite the limitations discussed in the previous sections, the application of nanoparticles is considered one of the most promising methods for the preservation of archaeological wood, especially considering that both actual and past consolidants for archaeological and waterlogged wood exhibit limitations such as long-term instability or even severe detrimental effects. This is the case, for instance, of the alum impregnation treatment used on wooden objects from the Viking Age Osberg find in Vestfold, Norway. During impregnation, the objects are heated in a supersaturated alum solution at temperature higher than 80 °C, and recent experiments indicated that sulfuric acid is released in the wood during the process (Braovac and Kutzke 2012). The resulting low pH (1.0–2.5) and the presence of metal ions such as Al, Fe, Cu and Zn lead to wood degradation. It has been advised that novel alternative consolidants should leave an open structure in wood, so to allow retreatment in the future if need be. This could be achieved either by foaming a polymer or by combining a polymer "spider web" with nanoparticles that help stabilize pH within the wood fibers (Christensen et al. 2012).

As a final example of the potentiality of nanoparticles dispersions, it is worth mentioning the treatment of the "Achromes", works made by the Conceptual artist Piero Manzoni using different materials and techniques, in response to Yves Klein's monochromes. In particular, Achromes made of cotton padding have been recently treated with calcium hydroxide nanoparticles to prevent acid hydrolysis and oxidation of cellulose.

5.4 Application Procedures

The aim of this section is to provide practical guidelines for the application of dispersions of $Ca(OH)_2$ (or $Mg(OH)_2$) nanoparticles in alcohols such as 1-propanol, 2-propanol and ethanol. In particular, these guidelines are to be used for the application of nanoparticles dispersions as designed by CSGI (Consorzio Interuniversitario per lo Sviluppo dei Sistemi a Grande Interfase, Center for Colloid and Surface Science), University of Florence, and registered as Nanorestore® and Nanorestore Paper®. Different formulations might behave differently, so adaptations of the indications here provided should be considered case by case.

Some of the applicative and safety rules described in Chap. 2 will be recalled as they apply also to the handling of dispersions for deacidification purposes. The proposed methods are based on the theoretical and practical knowledge gathered in the last decades through both research work and field activity in collaboration with conservators. It is fundamental to recall that conservation issues can vary greatly; therefore the practical guidelines reported here should be considered and applied critically, and not merely as recipes.

It is important to store the dispersions in firmly closed bottles, so to avoid any contact with air or moisture in order to prevent clustering of the nanoparticles (favored by water molecules) and uncontrolled transformation of $Ca(OH)_2$ into $CaCO_3$ (by reaction with CO_2). If stored in the above-mentioned conditions, Nanorestore® is a fairly stable dispersion but, especially at higher concentrations (5 g/L or more), deposits of nanoparticles can be observed in time. Thus, before use, the dispersion should be shaken energetically for a couple of minutes, and deposits need to be separated and discarded. If disposable, treatment with an ultrasonic bath can also be considered to favor the re-dispersion of particles' clusters. Both Nanorestore® and Nanorestore Paper® contain no additives, in addition Nanorestore Paper® has been formulated to avoid the sedimentation process, and is highly recommended for canvas and paper deacidification.

For paper and canvas deacidification, concentration of 2.5 g/L or less is typically advisable to grant acid neutralization and avoid white hazes on the substrate. Dilution of concentrated dispersions must be carried out using anhydrous solvents to prevent clustering or uncontrolled carbonation of the particles.

Before the application, tests on small spots must be carried out to check the sensitiveness of inks to the dispersion solvent.

The dispersion can be applied using different methods. Quicker treatments include spraying or brushing the dispersion directly onto the surface, or immersion of objects in the dispersion. Dripping the dispersion with a pipette allows better control of the applied quantity. In general, for a given volume of nanodispersion, the application must be carried out progressively, applying the dispersion up to saturation of the substrate and letting dry (until evaporation of the solvent) before applying further. Normally, for paper manuscripts and documents, half of the needed amount of dispersion for each paper sheet (see below) is applied on the front of the sheet, and the other half is applied on the backside of the sheet.

The alkaline content of paper is normally expressed as "alkalinity" in % $CaCO_3$ (m/m) or as "alkaline reserve" in mol kg^{-1}. The necessary quantity of alkalinity in the standard for permanent paper is generally set at 2% $CaCO_3$ (ca. 0.2 mol kg^{-1}, which represents ca. 1.7% $MgCO_3$), but this specification is arbitrary, and a higher alkaline content will provide a better protection from acidic gas pollutants and endogenous acids (Strlic and Kolar 2005). A precise determination of the acidity sources in the environment surrounding the artifacts, and the measurement of paper pH are helpful to evaluate the desired alkaline content that needs to be provided in order to counteract degradation both in the short and in the long term (when pH is ca. 5.5 or lower, paper is usually considered as acidic).

$Ca(OH)_2$ and/or $Mg(OH)_2$ nanoparticles can be applied to obtain a pH value (after carbonation of hydroxide) of 7.0–7.5, which grants protection against acidity and inhibits the catalytic oxidative activity of iron and copper ions. Too high pH values (e.g. 9 or higher) should be avoided on cellulose, in particular when oxidized, to prevent alkaline degradation processes such as β–alkoxy elimination.

The carbonation of calcium hydroxide nanoparticles is influenced by several parameters as explained in Chap. 2 (Sect. 2.3). A possible way to speed up the reaction is to keep the treated objects at RH values of 75% or higher, however it must be noticed that high RH values might favor the growth of microorganisms and other cellulose degradation mechanisms. Therefore, RH = 60% (at room T) can be considered as a reasonable condition for the storage of paper artifacts during the 10–15 days after treatment. Similar values (RH = 50 ± 5%, T = 18 ± 2 °C) can also be used as archival storage conditions, as indicated by Vosteen (1994). For instance, Poggi et al. (2014) reported that acidic paper samples (initial pH = 4.6) were treated with $Ca(OH)_2$ nanoparticles in ethanol to reach slightly basic values right after the treatment (pH = 8); after 10 days at RH = 50% and room T, pH had decreased down to 6.9 and remained such throughout more than 4 months (ca. 3400 h) of accelerated aging at RH = 75% and T = 80 °C. Analogous results were found for acidic canvas samples (initial pH = 3.7) treated with the same method.

In summary, the standard procedure for the application of nanoparticles involves the following main steps:

- First, measurement of pH is carried out. There are several methods for measuring the pH of paper, canvas and wood. Available and standardized methods for paper include cold extraction (TAPPI T 509 Om-02 2002; ASTM D778-97 2002) and determination of surface pH (TAPPI T529 Om-04, 2004). Typically, when pH is lower than 5.5, the deacidification treatment is deemed necessary.
- Theoretically, once the acid content of the artifact is known (e.g. as moles of H^+ ions or H_2SO_4 potentially released), it should be possible to calculate the amount of calcium hydroxide needed for deacidification, considering that one mole of $Ca(OH)_2$ (ca. 74 g) neutralizes 2 mols of H^+ ions or 1 mol of H_2SO_4. However, this evaluation is not feasible in many practical cases. Therefore we report below two standard applications, on paper at pH 4 and 5 respectively. For each case, we indicate the amount of nanodispersion (concentration = 1.5 g/L) needed either to reach neutralization or to obtain a slight excess of alkaline buffer (1–2% alkaline

reserve). The suggested values are based on real case studies, but the amounts needed in different situations might vary from case to case.

- **Applicationexample 1:** paper at pH 4 (paper grammage 100 g/m^2). To neutralize acidity, apply 8 mL (4 on the front, 4 on the backside) of nanodispersion (1.5 g/L) per each A4-size paper sheet. This corresponds to applying 128 mL per square meter (64 front, 64 backside). Therefore, using 1 L of nanodispersion it is possible to treat 7.8 m^2 of paper, or ca. 128 A4 sheets (including front and backside application). After full carbonation of $Ca(OH)_2$ (approx. 2 weeks after the treatment, keeping the treated object at RH=60% and room T), the final obtained pH is around 7 (to be checked through pH measurement).
- **To add an alkaline reserve (1–2%):** after the first treatment described above, apply another 25 mL (12.5 on the front, 12.5 on the backside) of nanodispersion (1.5 g/L) per each A4-size paper sheet. This corresponds to applying another 400 mL/m^2 (200 front, 200 backside). Therefore, using another liter of nanodispersion it is possible to treat 2.5 m^2 of paper, or ca. 40 A4 sheets (including front and backside application). After full carbonation of $Ca(OH)_2$ (approx. 2 weeks after the second treatment, keeping the treated object at RH=60% and room T), the final obtained pH is around 8.5 (to be checked through pH measurement).
- **Applicationexample 2:** paper at pH 5 (paper grammage 80 g/m^2). To neutralize acidity, apply 5 mL (2.5 on the front, 2.5 on the backside) of nanodispersion (1.5 g/L) per each A4-size paper sheet. This corresponds to applying 85 mL/m^2 (42.5 front, 42.5 backside). Therefore, using 1 L of nanodispersion it is possible to treat 12 m^2 of paper, or ca. 192 A4 sheets (including front and backside application). After full carbonation of $Ca(OH)_2$ (approx. 2 weeks after the treatment, keeping the treated object at RH=60% and room T), the final obtained pH is around 7 (to be checked through pH measurement).
- **To add an alkaline reserve (1–2%):** after the first treatment described above, apply another 25 mL (12.5 on the front, 12.5 on the backside) of nanodispersion (1.5 g/L) per each A4-size paper sheet. This corresponds to applying another 400 mL per square meter (200 front, 200 backside). Therefore, using another liter of nanodispersion it is possible to treat 2.5 m^2 of paper, or ca. 40 A4 sheets (including front and backside application). After full carbonation of $Ca(OH)_2$ (approx. 2 weeks after the second treatment, keeping the treated object at RH=60% and room T), the final obtained pH is around 8.5 (to be checked through pH measurement).
- In general, if the final pH (approx. 2 weeks after the treatment, keeping the treated object at RH=60% and room T) is lower than 7, the nanodispersion must be applied again until the desired pH value is obtained, in order to reach neutralization or to add the desired alkaline reserve.

The main applicative procedure for the treatment of paper is resumed in Fig. 5.10.

The spraying method is generally the most feasible for the backside of canvas paintings, which can be laid face down or kept vertical. In the first case it is easier to achieve a homogeneous treatment over the canvas surface, but the penetration of the dispersion through the painting layers might be enhanced.

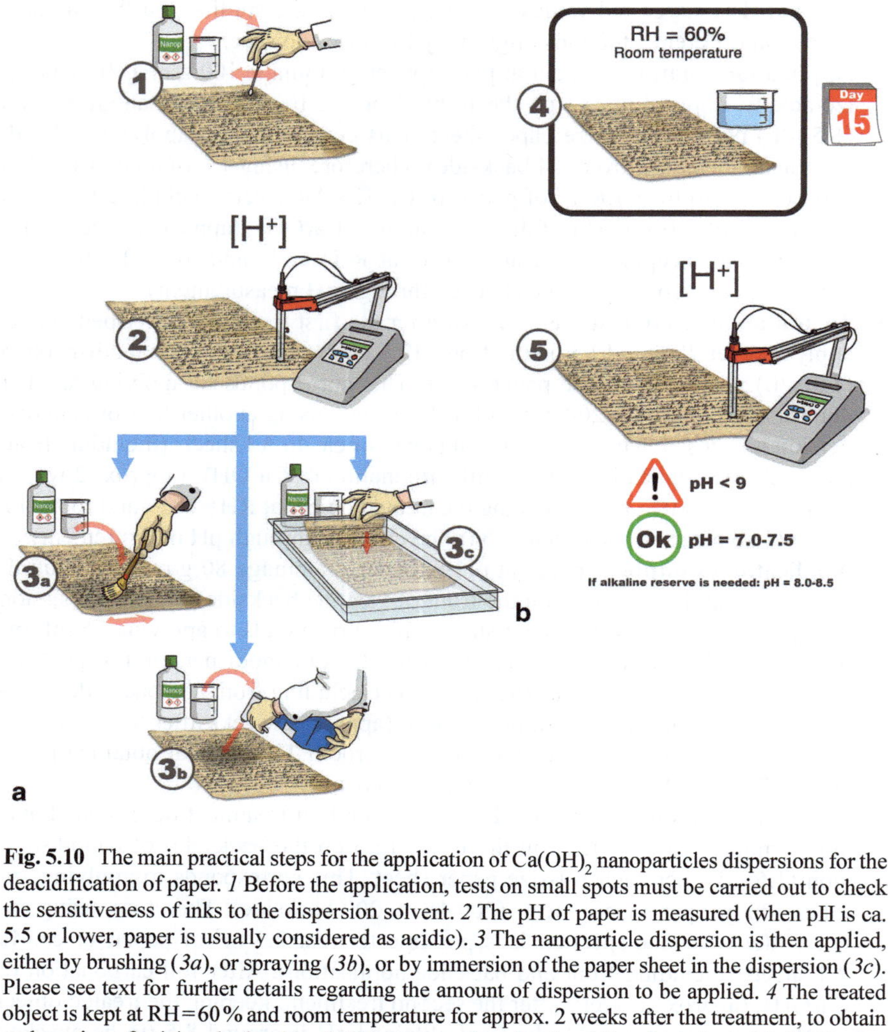

Fig. 5.10 The main practical steps for the application of $Ca(OH)_2$ nanoparticles dispersions for the deacidification of paper. *1* Before the application, tests on small spots must be carried out to check the sensitiveness of inks to the dispersion solvent. *2* The pH of paper is measured (when pH is ca. 5.5 or lower, paper is usually considered as acidic). *3* The nanoparticle dispersion is then applied, either by brushing (*3a*), or spraying (*3b*), or by immersion of the paper sheet in the dispersion (*3c*). Please see text for further details regarding the amount of dispersion to be applied. *4* The treated object is kept at RH = 60 % and room temperature for approx. 2 weeks after the treatment, to obtain carbonation of calcium hydroxide. *5* pH is measured again after full carbonation of $Ca(OH)_2$. Too high pH values (e.g. 9 or higher) should be avoided, to prevent alkaline degradation of cellulose. The final pH obtained should be around neutrality, or ca. 8.0–8.5 if an alkaline reserve is needed (see text for further details). (Image by Michele Baglioni)

Similar methods can be used for the treatment of wood surfaces such as boxes, cabinets or in general acid-emitting substrates.

So far, treatment of waterlogged wood samples (roughly 20 × 5 × 5 cm or smaller) has been carried out by immersion, but the method is still experimental. Application under vacuum or moderate pressurization, if judged feasible based on the object mechanical properties, enhances the penetration of particles greatly. As a preliminary step, the samples can be washed with water to partially remove the aged PEG. In case of significant PEG removal, re-treatment with fresh PEG or alternative consolidants might be carried out to prevent shrinkage.

The safety rules for the handling of dispersions of calcium hydroxide nanoparticles in short alcohols have been described in Chap. 2. However we resume here the main considerations. The ecotoxicological impact of the $Ca(OH)_2$ and $Mg(OH)_2$ nanoparticle formulations described in this section can be considered as similar to that of the dispersing media (e.g. 1-propanol, 2-propanol, ethanol), determining the same human health and environmental hazards of the organic solvents employed. Therefore the safe handling of the formulations involves the presence of good air ventilation, the use of suitable eye protection (lab glasses tested and approved under appropriate government standards, such as EN 166), no heating, and working without sparks, open flames and hot surfaces. Handling also involves the use of hand protection (nitrile rubber gloves, minimum layer thickness: 0.2 mm), even if Skin Irritation Tests confirmed that the formulations are not irritant for skin.

References

Afsharpour M, Hadadi M (2013) Titanium dioxide thin film: environmental control for preservation of paper-art-works. J Cult Herit. http://dx.doi.org/10.1016/j.culher.2013.10.008. Accessed 13 Oct 2014.

Anon (1988) Book preservation technologies, Congress of US, Office of Technology Assessment, Washington OTA-O-375, Washington, DC: U.S. Government Printing Office, May 1988.

ASTM D778–97 (2002) Standard test methods for hydrogen ion concentration (pH) of paper extracts (hot- extraction and cold-extraction procedures)

Baglioni P, Dei L, Giorgi R, Schettino C (2002) Basic suspensions: their preparation and use in processes for paper deacidification. International patent pending PCT/EP02/00319

Baglioni P, Chelazzi D, Giorgi R et al (2012) Nanoparticles for the conservation of cultural heritage: paper and wood. In: Somasundaran P (ed) Encyclopedia of surface and colloid science. Taylor & Francis, New York, pp 1–16

Banait NS, Jencks WP (1991) Reactions of anionic nucleophiles with α-D-glycopyranosyl fluoride in aqueous solution through a concerted, ADND (S_N2) mechanism. J Am Chem Soc 113:7951–7958

Banty JW, Maitland CL, Minter W et al (2010) Deacidification for the conservation and preservation of paper-based works: a review. Bioresources 5:1955–2023

Barrow WJ, Sproull RC (1959) Permanence in book papers. Science 129:1075–1084

Blüher A, Grossenbacher G (eds) (2006) Save paper! Mass deacidification, today's experiences, tomorrow's perspectives. Papers presented at the international conference, Swiss National Library, Bern, 15–17 Feb 2006

Boone T, Kidder L, Russick S (1998) Bookkeeper® for spray use in single item treatments. The Book and Paper Group Annual, vol 17. The American Institute for Conservation. http://cool.conservation-us.org/coolaic/sg/bpg/annual/v17/bp17-04.html. Accessed 14 May 2014.

Braovac S, Kutzke H (2012) The presence of sulfuric acid in alum-conserved wood—origin and consequences. J Cult Herit 13(3):S203–S208

Brewis S, Halsall TG (1961) The chemistry of triterpenes and related compounds. Part XXXVIII.* The acidic constituens of dammar resin. J Chem Soc 646–650

Bukovský V (2000) The influence of light on ageing of newsprint paper. Restaurator 21:55–76

Burgess HD, Kaminska EM, Boronyak-Szaplonczay A (1992) Evaluation of commercial mass deacidification processes: Akzo-DEZ, Wei T'o and fmc-mg3. Phase I: naturally aged papers. Canadian Conservation Institute, Ottawa

Calvini P (2005) The influence of leveling-off degree of polymerization on the kinetics of cellulose degradation. Cellulose 12:445–447

Calvini P, Grosso V, Hey M et al (1988) Deacidification of paper—a more fundamentl approach. Pap Conserv 12:35–39

Calvini P, Gorassini A, Merlani AL (2008) On the kinetics of cellulose degradation: looking beyond the pseudo zero order rate equation. Cellulose 15:193–203

Carter HA (1989) Chemistry in the comics. Part 3. The acidity of Paper. J Chem Educ 11:883–886

Carter H, Bégin, Grattan D (2000) Migration of volatile compounds through stacked sheets of paper during accelerated aging. Part 1: acid migration at 90 °C. Restaurator 21:77–84

Chadwick AV, Schofield E, Mark Jones A et al (2012) Ionic nanoparticles in heritage conservation: treaments for the Mary Rose timbers. Soli State Ionics 225:742–746

Chadwick AV, Howland K, Went MJ et al (2014) The application of ionic nanoparticles in the conservation of archaeological wood. Macromol Symp 337:74–79

Chelazzi D, Giorgi R, Baglioni P (2006) Nanotechnology for Vasa wood de-acidification. Macromol Symp 238:30–36

Christensen M, Kutske H, Hansen FK (2012) New materials used for the consolidation of archaeological wood—past attempts, present struggles, and future requirements. J Cult Herit 13S:183–S190

Dobrodskaya TV, Egoyants PA, Ikonnikov VK et al (2004) Treatment of paper with basic agents in alcohols and supercritical carbon dioxide to neutralize acid and prolong storage time. Russ J Appl Chem 77:2017–2021

El-Feky O, Hassan EA, Fadel SM et al (2014) Use of ZnO nanoparticles for protecting oil paintings on paper support against dirt, fungal attack, and UV aging. J Cult Herit 15:1665–172

Fors Y, Richards V (2010) The effects of the ammonia neutralizing treatment on marine archaeological Vasa wood. Stud Conserv 55:41–54

Foster G, Odlyha M, Hackney S (1997) Evaluation of the effects of environmental conditions and preventive conservation treatment on painting canvases. Thermochim Acta 294:81–89

Giorgi R, Dei L, Ceccato M et al (2002a) Nanotechnologies for conservation of cultural heritage: paper and canvas deacidification. Langmuir 18:8198–8203

Giorgi R, Dei L, Schettino C et al (2002b) A new method for paper deacidification based on calcium hydroxide dispersed in nonaqueous media. In: Daniels V, Donnithorne A, Smith P (eds) Preprint of IIC Baltimore Congress 2002 on works of art on paper, books, documents and photographs: techniques and conservation, Blatimore, (Sept 1–6, 2002). International Institute for Conservation, London, p 69

Giorgi R, Chelazzi D, Baglioni P (2005a) Nanoparticles of calcium hydroxide for wood conservation. The deacidification of the Vasa Warship. Langmuir 21:10743–10748

Giorgi R, Bozzi C, Dei L et al (2005b) Nanoparticles of $Mg(OH)_2$: synthesis and application to paper conservation. Langmuir 21:8495–8501

Giorgi R, Chelazzi D, Baglioni P (2006) Conservation of acid waterlogges shipwrecks: nanotechnologies for de-acidification. Appl Phys A 83:567–571

Giorgi R, Chelazzi D, Fratini E et al (2009) Nanoparticles of calcium hydroxide for wood deacidification: decreasing the emissions of organic acid vapors in church organ environments. J Cult Herit 10:206–213

Giorgi R (2013) in: Baglioni P, Chelazzi D (eds), Nanoscience for the conservation of works of art. The Royal Society of Chemistry, Cambridge, p 421–422

Glasser WG, Atalla RH, Blackwell J et al (2012) About the structure of cellulose: debating the Lindman hypothesis. Cellulose 19:589–598

Harris JF (1975) Acid hydrolysis and dehydration reactions of utilizing plant carbohydrates. App Polym Symp 28:131–144

Harris KE, Shahani CJ (1994) Mass deacidification: an initiative to refine the diethyl zinc process. http://cool.conservation-us.org/byorg/lc/massdeac/. Accessed 5 May 2014

Henniges U, Potthast A (2008) Phytate treatment of metallo-gallate inks: investigation of its effectiveness on model and historic paper samples. Restaurator 29:219–234

Henniges U, Banik G, Potthast A (2006) Comparison of aqueous and non-aqueous treatments of cellulose to reduce copper-catalyzed oxidation processes. Macromol Symp 232:129–136

Hubbe MA (2005) Acidic and alkaline sizings for printing, writing, and drawing papers. Book Pap Group Ann 23:139–151
Kolar J, Mozir A, Strlic M et al. (2007) Stabilisation of iron gall ink: aqueous treatment with magnesium phytate. E-Pres Sci 4:19–24
Kolar J, Mozir A, Balazic A et al (2008) New antioxidants for treatment of transition metal containing inks and pigments. Restaurator 29:184–198
Lòpez-Aparicio S, GrØntoft T, Odlyha M et al (2010) Measurement of organic and inorganic pollutants in microclimate frames for paintings. E-Pres Sci 7:59–70
Lundgaard LE, Hansen W, Linhjell D et al (2004) Ageing of oil-impregnated paper in power transformers. IEEE T Power Deliv 19:230–239
Mahltig B, Swaboda C, Roessler A et al (2008) Functionalising wood by nanosol application. J Mater Chem 18:3180–3192
Malesic J, Kolar J, Strlic M et al (2005) The use of halides for stabilization of iron gall ink containing paper—the pronounced effect of cation. E-Pres Sci 2:13–18
Medronho B, Romano A, Miguel MG et al (2012) Rationalizing cellulose (in)solubility: reviewing basic physico–chemical aspects and role of hydrophobic interactions. Cellulose 19:581–587
Neveel JG, Reissland B (1997) in Proc. Workshop on Iron Gall Ink Corrosion, Van der Windth H (ed), Amsterdam, p 37–46
Neevel JG (1999) The development of a new consveration treatment for ink corrosion, based on the natural anti-oxidant phytate. Int J Forensic Doc Examiners 5:130–137
Pauk S (1996) Bookkeeper mass deacidification process—some effects on twentieth-century library material. Abbey Newsl 20:50–53
Persson H, Turk M, Nyman M et al. (1998) Binding of Cu^{2+}, Zn^{2+}, and Cd^{2+} to Inositol Tri^{-}, $Tetra^{-}$, $Penta^{-}$, and Hexaphosphates. J Agric Food Chem 46:3194–3200
Poggi G, Giorgi R, Toccafondi N et al (2010) Hydroxide nanoparticles for deacidification and concomitant inhibition of iron-gall ink corrosion of paper. Langmuir 26(24):19084–19090
Poggi G, Baglioni P, Giorgi R (2011) Alkaline earth hydroxide nanoparticles for the inhibition of metal gall ink corrosion. Restaurator 32:247–273
Poggi G, Toccafondi N, Melita LN et al (2014) Calcium hydroxide nanoparticles for the conservation of Cultural Heritage: new formulations for the deacidification of cellulose-based artifacts. Appl Phys A 114:685–693
Santucci L, Plossi ZappalÃ (2001) Cellulose viscometric oxidometry. Restauratior 22:51–65
Schieweck A, Markewitz D, Salthammer T (2007) Screening emission analysis of construction materials and evaluation of airborne pollutants in newly constructed display cases. In Padfield T, Bochersen K (eds) Museum microclimates: conference on preventive conservation, 19–23 November 2007. National Museum of Denmark, Copenhagen, ISBN: 978-87-7602-080-4, p 67
Schofield E, Sarangi R, Mehta A et al (2011) Nnaoparticle de-acidification of the Mary Rose. Mater Today 14(7–8):354–358
Šelih VS, Strlic M, Kolar J et al (2007) The role of transition metals in oxidative degradation of cellulose. Polym Degrad Stabil 92:1476–1481
Selli E, Langè E, Mossa A et al (2000) Preservation treatments of aged papers by supercritical carbon dioxide. Macromol Mater Eng 280/281:71–75
Sequeira S, Casanova C, Cabrita EJ (2006) Deacidification of paper using dispersions of $Ca(OH)_2$ nanoparticles in isopropanol. Study of efficiency. J Cult Herit 7:264–272
Shanani CJ, Harrison G (2002) Spontaneous formation of acids in the natural aging of paper. In: Daniels V, Donnithorne A, Smith P (eds) Works of art on paper: books, documents and photographs: techniques and conservation. Contributions to the Baltimore Congress, 2–6 September 2002. IIC, pp 189–200
Stauderman SD, Brückle I, Bischoff JJ (1996) Observations on the use of Bookkeeper® deacidification spray for the treatment of individual objects. The Book and Paper Group Annual, vol 15, The American Institute for Conservation. http://cool.conservation-us.org/coolaic/sg/bpg/annual/v15/bp15-17.html. Accessed 14 May 2014
Stefanis E, Panayiotou C (2007) Protection of lignocellulosic and cellulosic paper by deacidification with dispersions of micro- and nano-particles of $Ca(OH)_2$ and $Mg(OH)_2$ in alcohols. Restaurator 28:185–200

Stefanis E, Panayiotou C (2008) Study of the photochelical stability of paper deacidified with dispersions of $Ca(OH)_2$ and $Mg(OH)_2$ nanoparticles in alcohols. Restaurator 29:125–138
Stefanis E, Panayiotou C (2010) Deacidification of documents containing iron gall ink with dispersions of $Ca(OH)_2$ and $Mg(OH)_2$ nanoparticles. Restaurator 31:19–40
Strlic M, Kolar J (eds) (2005) Ageing and stabilization of paper. National and University Library, Ljubljana
Strlic M, Kolar J, Šelih VS et al (2003) A comparative study of several transition metals in Fenton-like reaction systems at circum-neutral pH. Acta Chim Slov 50:619–632
TAPPI T 509 Om-02 (2002). Hydrogen ion concentration (pH) of paper extracts (cold extraction method)
TAPPI T 529 Om-04 (2004): surface pH measurement of paper
Tétreault J, Stamatopoulou E (1997) Determination of concentrations of acetic acid emitted from wood coatings in enclosures. Stud Conserv 42(3):141–156
Vosteen R (1994) Adviesrichtlijn luchtkwaliteit archieven, Ministerie van Volkshuisvesting, Ruimtelijke Ordening en Milieubeheer, Rijksgebouwendienst
Wouters J (2008) Coming soon to a library near you? Science 322:1196–1198
Yamjuan W, Yanxiong F, Wei T et al (2013) Preservation of aged paper using borax in alcohols and the superciritcal carbon dioxide system. J Cult Herit 14:16–22
Zhang Y, Bommuswamy J, Sinnot ML (1994) Kinetic isotope effect study of transition states for the hydrolyses of α- and β-glucopyranosyl fluorides. J Am Chem Soc 116:7557–7563
Zumbül S, Wuelfert S (2001) Chemical aspects of the Bookkeeper deacidification of cellulosic materials: the influence of surfactants. Stud Conserv 46:169–180

MIX
Papier aus verantwortungsvollen Quellen
Paper from responsible sources
FSC® C105338
FSC
www.fsc.org

If you have any concerns about our products,
you can contact us on
ProductSafety@springernature.com

In case Publisher is established outside the EU,
the EU authorized representative is:
Springer Nature Customer Service Center GmbH
Europaplatz 3, 69115 Heidelberg, Germany

Printed by Libri Plureos GmbH
in Hamburg, Germany